The Nature and Growth of
Ideas in Mathematics

Second Edition

Daniela Monaldi
York University

Kendall Hunt
publishing company

Cover image © bpk, Berlin/Gemaeldegalerie Alte Meister, Staatliche Kunstsammlungen, Dresden, Germany/Hans-Peter Klut. Inv.: Gal. Nr. 692./Art Resource, NY.

Kendall Hunt
publishing company

www.kendallhunt.com
Send all inquiries to:
4050 Westmark Drive
Dubuque, IA 52004-1840

ISBN 978-1-4652-5621-8

Printed in the United States of America
10 9 8 7 6 5 4 3 2 1

Contents

Chapter 1 The Beginnings of Numeracy 1

1.1 Counting Things 2

1.2 Writing Numbers 10

Chapter 2 The Evolution of Numbers 29

2.1 Broadening the Concept of Numbers 29

2.2 Developing the Algebraic Symbolism 38

Chapter 3 What Do We Know about Egyptian Mathematics? 45

3.1 The Primary Sources of Egyptian Mathematics 45

3.2 The Mathematical Papyri 47

3.3 Scribal Training, Professionalism, and the Practical Character of Egyptian Math 49

3.4 Mathematical Form and Content, and the Problem of "Translations" 51

3.5 Early Calculations of the Area of a Circle and Some Warnings on the "History of π" 53

Chapter 4 Greek Mathematics 59

4.1 The Sources of Greek Mathematics 59

4.2 Practical and Theoretical Mathematics in Ancient Greece 60

4.3 The Life of the Polis, the Habit of Debate, and the Requirement of Proof 66

4.4 Pythagoras and the so-called Pythagorean Theorem: Busting the Myth 71

4.5 The Discovery of Incommensurability 80
4.6 Hippocrates of Chios, the First *Elements of Geometry*, and the Squaring of the Circle 85
4.7 Archytas of Tarentum and the Duplication of the Cube 86
4.8 Plato and Mathematics 87

Chapter 5 Hellenistic Mathematics, Euclid's *The Elements* in Particular 91

5.1 The Golden Age of Greek Mathematics 91
5.2 Who was Euclid, and What Exactly Did He Write? 93
5.3 Euclid's *Elements* 96

Chapter 6 Mathematics and the Physical World in Antiquity 107

6.1 Mathematics and the World of Experience in Ancient Greece 107
6.2 Astronomy, Celestial Phenomena, and "Plato's Problem" 109
6.3 Hellenistic Astronomy, Claudius Ptolemy, and the *Almagest* 116
6.4 Saving the Phenomena 121

Chapter 7 Mathematics and the Origin of Modern Science Part 1 125

7.1 The Copernican Revolution, the Scientific Revolution, and the Birth of Modern Science 125
7.2 The Place of Mathematics in the Early Universities 127
7.3 The Christian-Ptolemaic-Aristotelian Cosmos 129
7.4 Copernicus's Heliocentric Hypothesis and the Role of Mathematical Hypotheses in Astronomy 132
7.5 Kepler and the Demise of the Celestial Spheres 139

Chapter 8 Mathematics and the Origins of Modern Science Part 2 149

8.1 The Trial of Galileo, or "How to Make the Earth Move without a Thousand Inconveniences" 149
8.2 Newton and the Mathematical Principles of Natural Philosophy 161

Chapter 9 From the Mathematics of Chances to the Theory of Probability 171

9.1 The Mathematical Theory of Probability 171
9.2 The Calculus of Chances 173
9.3 The Classical Theory of Probability 184
9.4 From Classical to Statistical Probability 199

Appendix 201

Chapter 1

The Beginnings of Numeracy

What are numbers? Every human society uses numerical concepts, yet different cultures handle them in different ways. Are numbers inherent in our minds or are they cultural constructs? They appear to be both. On the one hand, the work of cognitive scientists, linguists, and experimental psychologists provides evidence that certain numerical faculties are innate. For example, all humans, regardless of language and culture, have the same facility in distinguishing a group of four objects from a group of five and the same difficulty in telling nineteen objects from twenty; this suggests that there is a core set of numerical faculties that is common to all humans and thus, presumably, hard-wired in the mind. On the other hand, the work of historians and ethnologists brings to light striking differences among different cultures and between different historical periods. The ways in which people represent, communicate, and work with numbers are varied. The interrelations between numerical concepts and their forms of expression and representation—material, verbal, or written—are the subject of ongoing investigations on the part of researchers from several disciplines interested in the nature and uses of mathematics. One way to approach the question is to examine how the uses and expressions of numbers have evolved historically in relation to their social and cultural contexts.

1.1 Counting Things

Tallying

The oldest way to handle numbers is the use of notched sticks, knotted strings, and other artifacts that display a one-to-one correspondence between the markers (notches, knots, or other) and the objects to be counted. This way of recording discrete quantities is called tallying. Related artifacts, called "tallies", are found worldwide, especially in Africa, and date from at least the Upper Paleolithic era (35000–10000 BCE), as is shown by archaeological finds such as the Ishango bone (fig. 1.1).

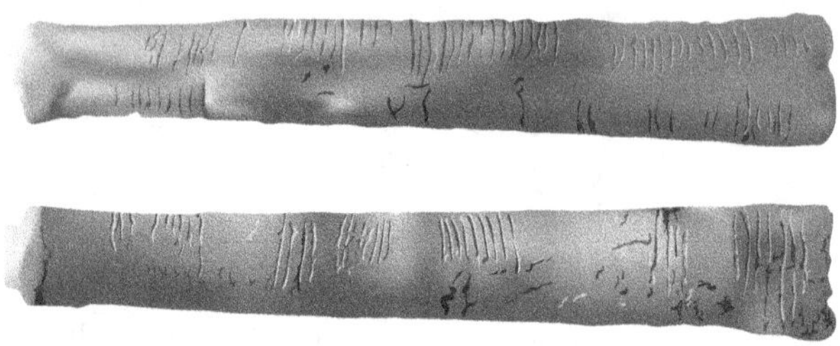

Fig. 1.1. A representation of the Ishango bone, one of the earliest known examples of a tally stick, probably a lunar calendar. See also http://www.math.buffalo.edu/mad/Ancient-Africa/ishango.html

The primary function of tallying is to record serially an ongoing enumeration, which is different from representing a final count or enabling arithmetic calculation. Even today, serial count is often taken by marking tallies in groups of five, with the fifth mark crossing the previous four. Tallying is also the precursor of written numeration, as is shown for example by the Etruscan-Roman number signs, I, II, III, V and X, which derive, to all evidence, from tally marks.

Tallies also evolved into implements for arithmetic calculations and accounting. For instance, "double tallies" have been used in many parts of the world for keeping record of deliveries, labour, commercial and financial transactions. A famous example is that of the British Exchequer tallies (fig.1.2).

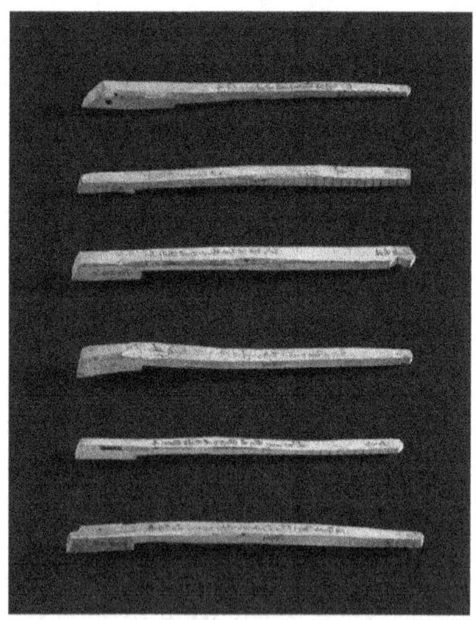

Fig. 1.2. British Exchequer tallies from the 1200s.

When a taxpayer owed or paid money up the Royal Treasury (called the Exchequer after a chequered counting board that was used for calculations) an official "Cutter" recorded the amount in the form of tallies on a wooden stick. The stick was then split lengthwise into two pieces, one of which (the *stock*) was kept by the payer and the other (the *foil*) by the Treasury, so that each party had an equal, legally binding record of the transaction. The Bank of England also used this method for loans and credits; stocks came to be used as money orders, and their owners were called *stockholders*. This method of accounting remained in use from the thirteenth century up to the 1820s. When the old store of tallies was destroyed in 1834, the furnaces used for burning became overloaded and set the Houses of Parliament alight.

The *Quipu*

The use of knots for tallying and data recording is widespread across cultures and ages. For example, the Greek historian Herodotus relates that King Darius of Persia, leaving for a doomed expedition into Scythia, gave his generals a leather strip on which he had tied sixty knots and ordered them to untie a knot each day as a means to keep count of time (Herodotus, *Histories* 4.98). Knots can be used also for more complex systems of information. An example is provided by early twentieth-century German millers, who closed their bags with different knots to alert the bakers not only to the quantity but also to the kind of flour the bags contained. The most sophisticated use of knots known is that of the *quipu* (or *khipu*).

Quipus are knotted-string artifacts that were made and used by the Incas, a people that flourished from around 1250 to 1560 in a region of the Andes comprising modern-day Peru and parts of Ecuador, Bolivia, Chile, and Argentina. Only about 600 quipus are preserved today. They are made of cotton or llama fibers, and are formed by a primary cord, usually about half-centimeter in diameter, to which a variable number of thinner strings, called pendant cords, are attached. A quipu can contain up to 1500 pendant cords. Some of the pendants have subsidiary strings attached to them, and subsidiaries to the subsidiaries; there are examples of quipus that contain up to thirteen levels of subsidiaries. Knots are tied into the pendants and the subsidiaries, forming a highly structured system of information (fig. 1.3).

Fig. 1.3. A quipu kept at the Museum of World Culture in Göteborg, Sweden.

The knots are of different kinds and are typically clustered at different levels on each string. Each level represents a power of ten; thus, the knotted strings form a decimal place-value system (fig. 1.4).[1] There are also examples of non-decimal quipus that were perhaps used for narrative rather than quantitative information.

Experts today are able to interpret the quantitative data represented by the knots; however, they are unable to read the accompanying information encoded in the colours, twists, and other features, which probably indicate the kinds of items enumerated by the knots. To interpret the quipus fully we have to rely on the Spanish documents from the years following the European invasion of the Inca empire in the sixteenth century.

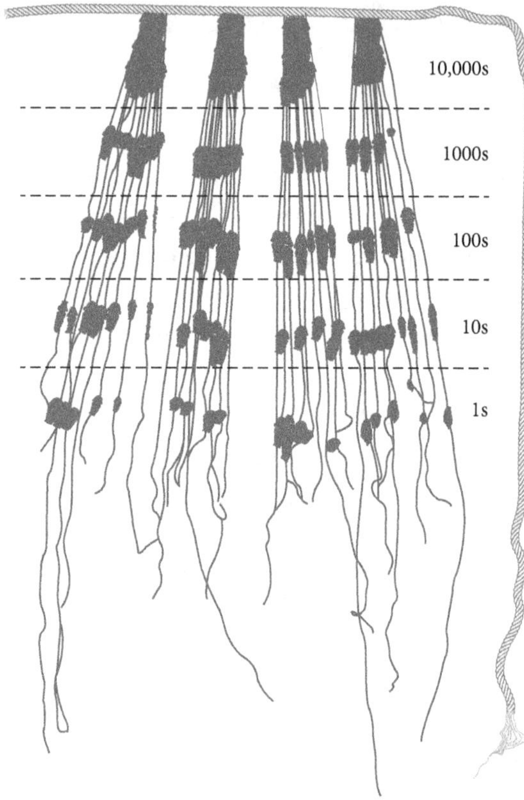

Fig. 1.4. A representation of knot clusters in a quipu, recording data according to a decimal place-value system.

[1] Place-value or positional systems are discussed in the next sections.

According to these documents, the quipus contained data of censuses, tributes, stored goods, astronomical and seasonal events related to the calendar, royal genealogies, historical facts, and so on. In other terms, the quipus served the same functions as the earliest examples of writing in Middle Eastern civilizations (and as the vast majority of written documents ever since): they were instruments of the administration of the state. The most prominent use was the implementation of a labour tax system, in which all men (some records say women also) between the ages of 18 and 50 would owe a certain amount of labour to the state. Detailed studies have shown that the quipu also encoded arithmetic operations, and possibly a method of checks and balance similar to double-entry bookkeeping. A special class of quipu-makers and keepers, called *quipucamayucs*, formed the bureaucracy of the Inca empire. These administrative officers were trained in a school set up in the Inca capital, Cuzco, where they learned the technical skills of the quipu, as well as language, religion, loyalty, and the principles of law and administration. In sum, despite different material and appearance, the technology of the quipu is remarkably similar to that of our written numerals. Like the latter, it formed the basis of a system of social, economic, and legal control, a system that was at least as complex and precise as that of the Europeans at the time of the conquest.

The Inca empire was vanquished by Francisco Pizarro and his small contingent of 164 conquistadores. The conquest began in 1532 with the massacre of Cajamarca and the capture, betrayal, and execution of Atahualpa, one of the two contenders in the succession to the Inca throne. Pizarro then led his troops to Cuzco and conquered it in 1534. There followed a long war between the Spanish invaders and the Inca resistance, which ended only in 1572 with the execution of the Inca leader, Tupac Amaru. In order to facilitate the normalization of the conquest and the pacification of the new colony, the Spanish viceroy Francisco de Toledo undertook a sweeping reorganization that transformed the Andean territory and society. The Toledan reforms included the founding of new cities and administrative districts, the displacement of communities, religious indoctrination, and the exaction of taxes in coins (a new concept for the Inca people) in addition to tributes in labour, agricultural produce, and manufactured goods. Essential to this process was the census, which the new Spanish administrators could only conduct successfully by enlisting the collaboration of local lords and quipucamayucs, and by launching a program of translation of the quipu data into written data.

In this way, the quipu-based numeracy and literacy of the conquered were displaced by the paper-based numeracy and literacy of the conquistadores. The Spaniards were initially respectful of the quipus and their keepers, as they had to rely on them to establish their colonial rule. While many native Andeans learned the alphabetical script and the western numerals, it appears that no Spaniard mastered in the quipu technique. As more and more data were transferred to the written texts, the quipus turned from needed tools to symbols of Inca autonomy and a threat to the authority of the colonial officials. By the end of the sixteenth century, the Spanish rulers declared them to be idolatrous, instruments of the devil; their use was prohibited and thousands of them were destroyed. Only a few hundred exemplars survived, although the technique continued to be used by villagers for local record-keeping purposes, in some cases to the present day.

The Counting Board

Across different ages and cultures, people have developed methods of calculation that were functional to the tasks at hand, supporting mental processes with suitable technologies. The most portable and inexpensive calculating tools are the fingers. Reckoning with fingers was common in ancient Greece and Rome as well as in the medieval Middle East and Europe, and it was widely used in trade and bargaining even between speakers of different languages. In fact, the term "digit" derives from the Latin word *digitus*, which means finger or toe. Another very common means of calculation was the counting board, a flat surface marked with lines, on which pebbles or counters were placed and moved to perform arithmetic operations. The ancient Greeks called the counting board *abákion*, which means "table without legs", and the Romans converted the term into *abacus*. The abacus technology also left its mark in our language: the verb "calculate" derives from the Latin *calculus*, which means small stone. The oldest Greek abacus surviving today is a marble tablet found on the island of Salamis.

The lines on the Salamis tablet are labeled with letters indicating the names of ancient Greek currency, showing that the calculations made with it were mainly for money transactions. Artistic depictions of ancient life scenes also show that the principal use of the abacus was for monetary calculations. For example, in a fourth-century vase on which King Darius of Persia and his court are painted, the royal treasurer is portrayed making use of a counting board to reckon the tributes brought to the king by people carrying sacks of coins (fig. 1.5).

Fig. 1.5. Detail of the Darius vase.

The Romans used abaci with loose counters similar to the Greek, as well as hand-held devices in which the counters slid into grooves (fig. 1.6).

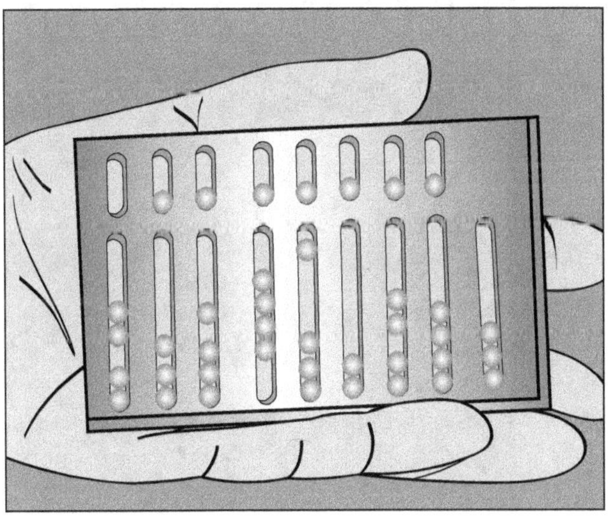

Fig. 1.6. A representation of a Roman hand-held abacus.

The use of counting boards on tables by merchants, moneylenders, and money exchangers in early-modern Italy, where such tables were called *banche*, gives us the term "bank". In East Asia, counting rods (*suan zi*) made of bamboo, wood, ivory, or bone, and arranged in columns were used as calculation devices up to the fifteenth century. Various types of counting boards with bead-like counters that run along wires fixed on a frame have remained in common use until recent times; examples are the Chinese *suan pan*, the Japanese *soroban*, and the Russian ščët. Each type of device uses a specific calculation technique.

In the modern era, in Europe and other parts of the world under European influence, the counting-board technology was replaced by the pen-and-paper technique based on the Hindu-Arabic numerals. This new form of numerals originated in India and spread throughout Europe from North Africa and Spain, which were part of the Islamic empire.[2] The textbooks that explained their use became known as *algorisms*, after al-Khwārizmī, the author of the Arabic book on which most of the Latin texts were based; the name was also applied to the method itself, and gave origin to the modern term "algorithm". Arithmetic textbooks became a popular and profitable genre due to the demand for calculation skills from the rising merchant class in the expanding money economy. The authors of these books, the *algorismists*, tended to promote the algorism over the abacus, often presenting their argument as a competition between the new and the old, as is seen for example in a famous illustration from a book titled *Margarita Philosophica*, published in 1508 by the German Gregor Reisch (fig. 1.7).

Fig. 1.7. The competition between the algorism and the abacus, from Gregor Reisch, *Margarita Philosophica*, 1508.

In Reisch's allegory, a female figure, personifying Arithmetic, oversees the competition between the two types of arithmetic; the new is championed by a fashionably-dressed young man named Boethius, after a Latin philosopher of the sixth century, to which the new numerals were mistakenly attributed, and the old by an elderly man named Pythagoras, after the ancient Greek philosopher mythically credited with the invention of the counting board.

[2] See next section.

The transition from the old to the new system was a slow process. It is popular belief that Hindu-Arabic numerals prevailed in early-modern Europe because they were technically superior, more suitable to calculation and more efficient than the old Roman numerals. Anybody schooled in the Hindu-Arabic system will be inclined to support this opinion: to carry out even the simplest arithmetic operations on paper with Roman numerals is a clumsy affair compared with the streamlined algorithm that uses Hindu-Arabic numerals. The same applies to the Greek system of numerals, which used the letters of the Greek alphabet in a way similar to the Roman system. Greek and Roman numerals, however, were never used to perform calculations. They were only used to write down the results obtained with the counting board. Their function was to represent and record numbers, while the function of supporting the arithmetical operations was left to the fingers or the abacus. In contrast, Hindu-Arabic numerals are both a system of representation and a technology of calculation; they are used at the same time to operate and to represent. Whether the written algorism is more efficient than the material abacus as a means of calculation is yet to be demonstrated. A skilled counting-board operator can perform complex arithmetic calculations surprisingly quickly and well. In 1946, a military magazine in Tokyo staged a competition between a young Japanese soroban master and an American military technician using one of the electric calculators then available. Although the match was clearly designed to display the superiority of the electric technology, the Japanese calculator won four events out of five.

The principle that gives the Hindu-Arabic numbers their calculation power is the same as the working principle of the abacus. It is the positional or place-value principle. According to this principle the digits acquire their values according to their position relative to the other digits in a number in the same way that counters change value according to the column of the counting board on which they are placed. The mental operations that are facilitated by the movements of the counters on the counting board closely resemble those performed when reckoning with written positional numerals. The following example illustrates the application of the positional principle with a counting board of the type shown in Reisch's picture (fig. 1.8). In the diagram on the left, two numbers, 3507 and 7249 in our decimal system, are represented by means of a place-value system in which the successive lines represent the powers of ten (the bottom line is for the units, the lines above for the tens and hundreds, the line marked "x" is for the thousands, etc.) and the spaces between the lines represent the intermediate values (equal to groupings of five relative to the line below). The diagram on the right represents the result of the addition of the two numbers, and the intermediate stages of the calculation are shown in the middle.

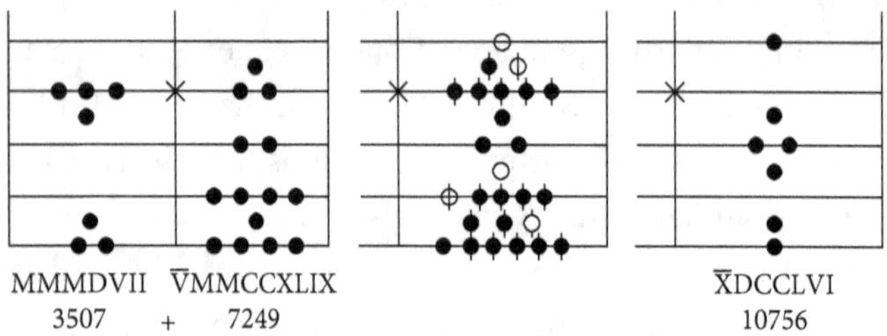

Fig. 1.8. The sum of two numbers performed on an early-modern European counting board. (Modified from Menninger, *Number Words and Number Symbols.*)

By the start of the seventeenth century, the Hindu-Arabic number system had become the system of choice of European mathematicians and it eventually came to dominate western mathematics. This may have occurred less because of its alleged technical superiority than for other factors that depended on social and cultural circumstances; these would include the appeal of novelty to the upwardly mobile classes of mercantile capitalism, the increasing availability of affordable ink and paper, and the adaptability of the system to the standardization of characters and page composition required by one of the seminal technologies of modernity, typography.

1.2 Writing Numbers

Systems of Numerals and the Positional Principle

While tallying is cross-cultural and predates writing, in some societies tallying evolved into numerals, that is, written signs representing numerical quantities. Numerals are, in fact, among the earliest forms of writing, possibly the very earliest. Starting with the Sumerian civilization at the end of the fourth millennium BCE, over a hundred different systems of numerals have been recorded around the world. They use sets of two to forty number signs, grouping numbers according to one or more number bases (commonly but not always five and ten) and combining them together to form every other number. Most systems are simply additive: the value of the number represented is given by the sum of the values of all the symbols used to represent it. The simplest additive system consists of tally marks. It uses a single sign, for example a scratch or a line, often supplemented by a transversal line to mark grouping of five or so. Many systems use special signs to signify different values. An example of an additive decimal system is given by the hieroglyphic numerals carved in stone in ceremonial buildings and artifacts by the ancient Egyptians (figs. 1.9 and 1.10).

Fig. 1.9. Ancient Egyptian hieroglyphic numerals.

Fig. 1.10. Hieroglyphic numbers inscribed on the ceremonial mace head of pharaoh Narmer, recording his war spoil (ca. 3000 BCE).

In such a system, the values of the numerals are fixed and independent of position and arrangement. Other systems give fixed values to the individual signs, but adopt specific orderings to form the overall number value. In the Roman system, for instance, the numerals I, V, X, L, C, M, form different values according to whether the sign of lower value is placed to the left or to the right of that of higher value (fig. 1.11).

Fig. 1.11. Example of Roman numerals.

While all systems of numerals have the purpose of representing numbers, some of them serve also as computation technologies. These systems are based on the same principle as the counting board, the positional or place-value principle.[3] In a positional system, although the number signs, or digits, have given primary values, in general their values are not

[3]The application of the positional principle in the abacus is explained in the previous section.

fixed but change according to their relative positions. A number is formed by ordering the number signs in ranks, with each rank having a place value equal to a power of the base of the system. When placed in the central rank, the number sign assumes its primary value. When placed in the higher ranks, it is multiplied by the successive powers of the base. When placed in the lower ranks, it is divided by the successive powers of the base. All of this is easily grasped by considering specific examples, such as our own system (decimal, or in base ten), the binary system familiar to computer scientists (in base two), the Babylonian system (sexagesimal, or in base sixty, see fig. 1.12 and fig. 1.13), or the Maya system (vigesimal, or in base twenty, see fig. 1.14).

The most ancient system of numerals has been found in clay tablets excavated in the Middle East and dated from the late fourth millennium BCE. In its earliest form, the representation consisted of imprints of clay tokens representing different kinds of countable items, and served the administrative needs of the temple-based economy in the city-state of Uruk and other Sumerian cities in the region known as Mesopotamia. The token imprints eventually evolved into cuneiform (i.e., wedge-shaped) signs made with a stylus. Archaeologists regard the transition from token imprints to number signs as a milestone in the process of the invention of writing. In addition, they have also interpreted it as a reflection of the conceptual leap from understanding numbers as attributes of concrete objects (e.g., three houses, three goats, three pebbles, etc.) to understanding numbers as independent abstract entities (e.g., the number three). In fact, while the tokens and the token imprints signified the counted objects, representing both the kind and the quantity of objects, the number signs signified solely the numerical quantities. By the beginning of the second millennium BCE, in the period that saw the rise to dominance of the city of Babylon (Old Babylonian, or OB, period), the professional class of the temple administrators,

Fig. 1.12. Old Babylonian tablet with cuneiform numbers (ca. 1800–1600 BCE). Yale Babylonian Collection.

the scribes, had developed a cuneiform script and a cuneiform system of numerals (fig. 1.12).

The Babylonian numerals consisted of two signs, the upright wedge and the sideways wedge, representing respectively the unit and ten units. These two signs were combined, in an additive and positional way, to form any number. They were added to one another to form the numbers from one to fifty-nine, and were arranged according to place values equal to the powers of sixty to represent the numbers upwards from sixty. In other terms, the Babylonian system of numerals was a mix of a decimal additive system and a sexagesimal positional system (fig. 1.13).

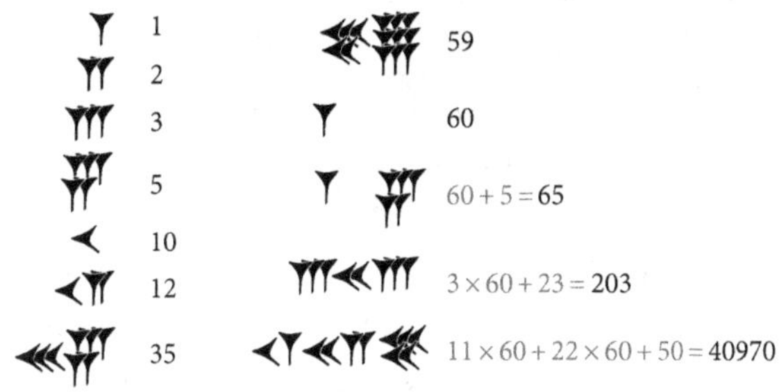

Fig. 1.13. Example of numbers written in the cuneiform sexagesimal system and in our decimal positional system. The Old Babylonian system of numerals did not include a symbol equivalent to our 0. The unit sign, for example, could signify 1, 60, 1/60, and other powers of 60, depending on the context of use. Only after 600 BCE did Babylonian astronomers begin to use a special sign to indicate an empty position within a number.

Besides the Babylonian system and the abacus, the positional principle was used in China from around 400 BCE (Chinese so-called rod-numerals, actually bamboo sticks used in a way similar to a counting board), in Mesoamerica from no later than 100 CE

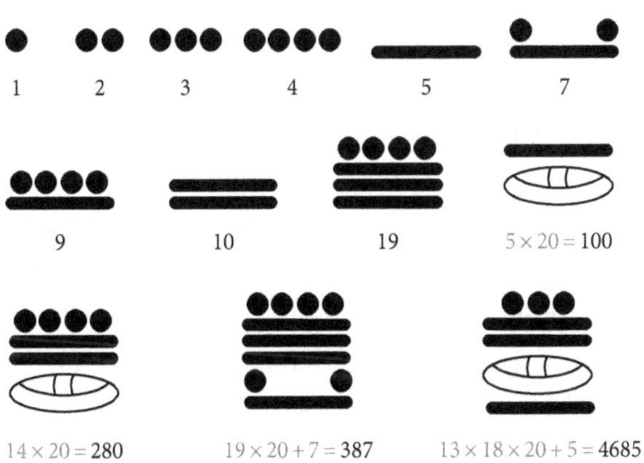

Fig. 1.14. Examples of Mayan numbers (from ca. 100 CE). The Maya used positional system of numerals that was mostly vigesimal (in base 20), with the exception of the third-position place value, which was 18. The system included a special symbol, an oval shape vaguely similar to an eye or a shell, which signified an empty place. See also fig. 1.16.

(Mayan numerals), in India from around 500 CE (Hindu numerals), and in the Andes region from around 1300 CE (Inca quipu knot-numerals). In some cases, it is possible to conjecture that trade exchanges between different populations transported the positional principle from one place to another, for example, from Mesopotamia to India, or from China to India in the form of the rod-numerals, and vice versa. These exchanges are very difficult to document, however, because there is no written evidence. The only documented case of diffusion is that of the Hindu numerals, which were brought from India to the Middle East and to Europe in the Middle Ages, as we shall see in the next section. In any case, there are interesting differences in the bases, the symbols, and the uses of the various positional systems. It is possible that positional numerals developed independently in different places at different times. The relation between the use of counting boards and the invention of written positional numerals needs more investigation. What is clear is that in the case of medieval Europe there was no simple transfer of the positional principle from the abacus to the written page.

The Diffusion of Hindu-Arabic Numerals

We do not know how and when exactly the positional system emerged in India. What seems clear is that by the beginning of the seventh century CE a written number system existed, which applied the positional principle to the first nine digits of an older, non-positional system, the Brahmi numerals. The nine Brahmi symbols were supplemented by a new sign, an empty circle, which signified an empty position within or at the end of a number (fig. 1.15).

Knowledge of the system spread quickly to the Middle East, as is testified by the words of a Syrian scholar, Severus Sebokht, who in 665 wrote admiringly of the mathematical subtlety of the Indians and their "clever method of calculation that surpasses all words,… that which is done with nine symbols." (Quoted in Burnett 2006, 15.) It is possible that the numerals traveled with traders, who used them in their dealings, but we have no documents to support this hypothesis. We know, however, that Indian mathematics and astronomy were brought to Baghdad by a diplomatic mission to the court of the caliph al-Mansour in 771. Several Arabic scholars learned the Indian system of numerals and started to write handbooks, typically titled *kitab al-hisab al-hindi* (*Book on the Indian Calculation*) to teach it. The first and most famous of these manuals was authored around 830 by the astronomer and mathematician Muhammad ibn-Musa al-Khwārizmī.

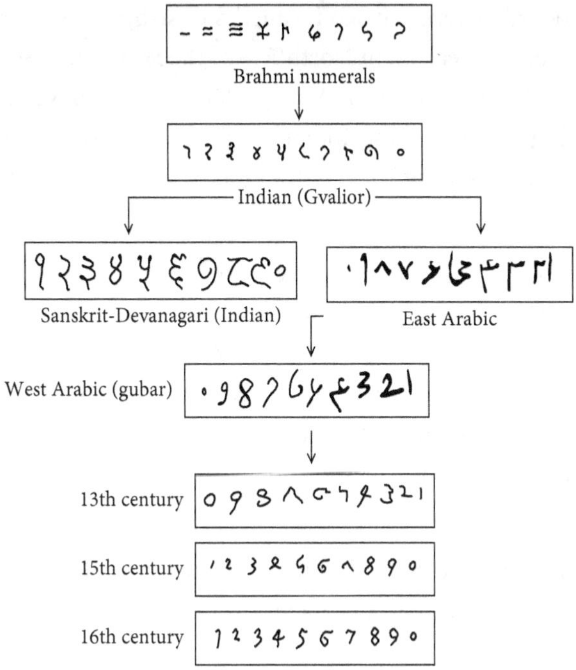

Fig. 1.15. The genealogic tree of the Hindu-Arabic numerals. (Modified from Menninger, *Number Words and Number Symbols.*)

The positional principle thus returned to the land of Mesopotamia in a new guise. By this time, however, the professional tradition of the ancient scribes had long ended, and all knowledge of their numerals was lost. Al-Khwārizmī and his contemporaries were unable to make any connection between the new Hindu system and the ancient cuneiform system. We have lost the Arabic original of this book but we do have several Latin versions of it (not literal translations), made in Spain and Italy in the twelfth century. Since typically these versions began with a reference to al-Khwārizmī's name, for example, with the formula "*Dixit Algorizmi*" ("al-Khwārizmī said"), the Hindu system became known in Europe as "algorism". Remarkably, although positional numerals can be used to represent both integer and fractional values, and indeed were so used by the Babylonians, the Hindu numerals were initially used for integer values only. It took several centuries of practice with this system for mathematicians in Islam and in Europe to develop a decimal notation for fractions, and to recognize the power of the incospicuous tenth digit, the empty circle.

The Hindy system did not take root easily in the Middle East and in Europe. Different Arabic authors expressed contrasting opinions about the method and how it compared to finger calculation. One problem was that the new method required writing implements.

The Arab calculators used a framed board covered in sand, called a dust-board, where they traced the numerals with a stick. For this reason, in North Africa the Hindu numerals became known as *gubar* or *ghobar*, which means dust. Around 860 a scholar advised schoolteachers to teach finger counting instead of the Indian method, because finger counting needed "neither spoken word nor writing". Similarly, another author wrote in 946 (quoted in Kunitzsch 2003, 5):

> The scribes in the administration refrain from using these [Indian] numerals because they require the use of materials, and they think that a system which calls for no materials and which a man can use without any instrument apart from one of his limbs is more appropriate in ensuring secrecy and more in keeping with their dignity; this system is computation with the joints and tips of the fingers, to which they restrict themselves.

On the other hand, around the same time the mathematician Abu al-Hasan al-Uqlīdisī ("the Euclidean") wrote an arithmetic textbook in which he explained both the traditional and the Hindu system and highlighted the advantages of the latter as follows (quoted in Katz 2009, 269):

> Most scribes will have to use [the Indian method] because it is easy, quick, and needs little precaution, little time to get the answer, and little keeping of the heart busy with the working that he has to see between his hands, to the extent that if he talks, that will not spoil his work; and if he leaves and busies himself with something else, when he turns back to it, he will find it the same and thus proceed, saving the trouble of memorizing it and keeping the heart busy with it. This is not the case in the other [method], which requires finger bending and other necessaries. Most calculators will have to use [the Indian method] with numbers that cannot be managed by the hand because they are big.

Al-Uqlīdisī was aware that some people thought it "ugly to see the [dust board] in the hands of the scribe ... sitting in the market places" (quoted in Katz 2009, 269); therefore, besides explaining the method, he taught how to use it with pen and paper, to replace the dust board with a more elegant, if more expensive, medium.

By the end of the twelfth century, the Hindu numerals had traveled across North Africa to Spain, and a "western" version of them had been developed in the multi-cultural city of Toledo. It is in this modified form that they became known to the Latin scholars, who travelled to the city to learn and translate the science of the Arabs. Since the exact shapes of numerals eventually adopted by Europeans and transmitted to us are found almost exclusively in Latin translations of Arab texts, some historians maintain—perhaps too fastidiously—that we should call them "western" rather than Hindu-Arabic.

The new numerals reached also non-academic users thanks to the mathematically talented son of an Italian merchant, Leonardo of Pisa, nicknamed Fibonacci. As a child, Leonardo was sent by his father to North Africa to be educated in the arts of calculation needed for the practice of commerce. He brought his knowledge back to Pisa and started a career as author of mathematical manuals for businessmen. The most famous of these is *Liber abaci* (*The Book of Calculation*), first published in 1202, in which Leonardo expounded the Hindu-Arabic system and demonstrated its use in a series of number problems. His explanation begins,

> The nine figures of the Indians are
>
> 9 8 7 6 5 4 3 2 1.
>
> With these nine figures, and with the sign 0 which the Arabs call zephyr, any number whatsoever is written, as is demonstrated below.

Fibonacci's *Liber abaci* initiated a whole genre of arithmetic handbooks. These texts were written on the model of al-Khwārizmī's *kitab al-hisab al-hindi* and were used in a new type of school that appeared in Europe between the fourteenth and the sixteenth centuries to teach the youth of the mercantile class the computational skills needed for careers in commerce and banking. These schools, called "abacus schools" (the term abacus is used in this case to mean calculation in general) responded to a rising demand for practical mathematical knowledge in European society. Spurred by the Crusades and by new technologies of navigation, a commercial revolution was underway in the Mediterranean and surrounding areas. Merchants no longer traveled over the seas with their cargos; they stayed at home and acted as shipping agents in a number of commercial ventures at the same time. International trading companies sprouted in seaport cities in Italy and other coastal regions. The growth of international commerce entailed frequent and often complex calculations of currency exchange. Business partnership required computations concerning the appropriate share of profits, and entrepreneurs began using financial instruments such as promissory notes and letters of credit. With money now used as capital for investments, interest on loans was no longer seen as morally objectionable, and compound interest calculations became a necessity of commercial life. Double-entry bookkeeping was introduced in Northern Italy in the fifteenth century and spread from there to the rest of Europe. While trade had always required basic calculation skills, the new money economy created a widespread need for more advanced forms of numeracy. The traditional academic institutions, the universities, were unable to meet this new demand. Their curriculum included mathematics only as one component of the Liberal Arts, which were taught at the lowest level as preparation for the higher studies of philosophy, theology, medicine, and law. Accordingly, university mathematics consisted for the most part of the basics of Greek theoretical arithmetic and geometry, disciplines concerned with abstract questions of a distinctively unpractical character, as for instance, "what is a number?"[4] For this reason, commercial mathematics, geared toward practical problem-solving and driven by an interest in what numbers could do, developed mostly in non-academic venues like the abacus schools.

The subtle and exotic Hindu system thrilled a few mathematical enthusiasts, but its diffusion was not a smooth ride. In Islam, scientists continued to use the traditional systems well into the thirteenth century. In Europe, it took four centuries for the algorism to go from Fibonacci's trail-blazing introduction to full endorsement by the mathematical community. As we have seen, it had to compete with trusted techniques like finger counting and the abacus. It also required writing implements, which were perceived as undignified like the dust board, or were costly and difficult to procure, like ink and paper or parchment before the sixteenth century. Furthermore, its unfamiliarity engendered mistrust. To many, it looked like a secret code that fraudsters could turn against ordinary people. In 1299 the city of Florence banned its use in bookkeeping on the grounds that the digits, which did not yet have standardized shapes, could be too easily misunderstood and falsified. A bookkeeping manual from Venice also warned that the "figures of the modern abacus" (i.e., the Hindu-Arabic digits) could be easily retouched; for instance, the "figure for nothing" could be turned into a six or a nine. Moreover, in 1348, the University of Padua required all booksellers to list their prices "*non per cifras sed per litteras clara*" ("not by figures but by clear letters", quoted in Swetz 1987, 182)—just as we do when writing a cheque.

In addition to practical inconveniences, the new system also presented conceptual obstacles. Its capacity to represent *any number whatsoever* with only nine digits bewildered the early users. The textbook authors stressed over and again that each digit

[4] Greek mathematics is the topic of Chapters 4 and 5.

did not correspond to a single value, as the Roman numerals did, but to an endless array of possible values. What determined the actual value of each digit in a specific number was the place it occupied relative to the other digits. Ordering the digits correctly was essential to the method, but the required order was confusing. When presenting the digits individually and in sequence, Indian and Arabic mathematicians followed the order of counting: 1, 2, 3, 4, 5, 6, 7, 8, 9. Arabic script proceeds from right to left. Latin script, in contrast, proceeds rightward. The Latin translators had to choose whether to keep the Arabic visual order or invert it. Fibonacci and his followers up to the sixteenth century opted for visual fidelity to their sources over numeric logic. More importantly, when the digits were assembled to form a number, they were arranged left to right according to decreasing values, similarly to the Greek and Latin numerals. In the case of the Hindu-Arabic system, however, this ordering was counterintuitive because it meant that the value of a digit was determined by the digits that followed it in the order of writing. Not only were the sign values decided only after the whole number was completed, but the decision was also made, so to speak, by the lowliest ranks. The lower ranks had the power to increase retroactively the value of the higher ones. To ease the discomfort, early algorismists instructed their readers to write the numbers "backward", that is, right to left, until they became proficient in using the method. The positional method seems natural to us because we have become familiar with it at an early age. At the end of the sixteenth century, however, it was still perceived as an odd conceit. William Shakespeare often made references to it in his plays. He opened *Henry V* with a comparison between the Hindu-Arabic numbers ("ciphers") and the illusion of theatre (Shakespeare, *Henry V*, "Prologue", quoted in Ostashevsky 2004, 212):

> *O pardon! Since a crooked figure may*
>
> *Attest in little place a million*
>
> *And let us, ciphers to this great accompt*
>
> *On your imaginary forces work.*

A few actors on a stage had the power to represent great historical events thanks to the "imaginary forces" of the spectator in the same way as a "crooked figure" (the zero), a mere sign having no value by itself, and what's more in a "little place", could conjure up a million thanks to the convention of the positional principle. Shakespeare used the writing of Hindu-Arabic numbers also as a metaphor for the claim of dynastic rights by Henry V, a claim that he defined as "sinister" (meaning leftward but also devious). Henry won the throne of England not because his ancestry made him the legitimate heir, but because he was the victor of the battle of Agincourt. Asserting his claim by force, Henry's final victory sanctioned his dynastic line retroactively, thus writing history, so to speak, backward.

The Problem of Zero

The puzzlement caused by the shifty ciphers and their sinister ordering was amplified by the ghostly character of the "crooked figure", a sign that did not stand for a quantity but for the absence of any quantity. Greek and Roman numerals did not include anything like it. Practical mathematicians working with tokens and counting boards have no need for a special sign to mark an empty column. Since they use numerals only to record quantities rather than carry out operations with them, they have no reason to invent a symbol that represents no quantity. The need for an empty-position sign arises, when it does, only in a written positional system, as is illustrated, for example, by the "shell glyph" of the Mayan system (fig. 1.16).

As we have seen, late Babylonian astronomers started using an empty-position sign sometime between 600 and 300 BCE. Shortly before 300 BCE, Alexander the Great conquered Mesopotamia and unified it with Greece in his large empire. Greek astronomers

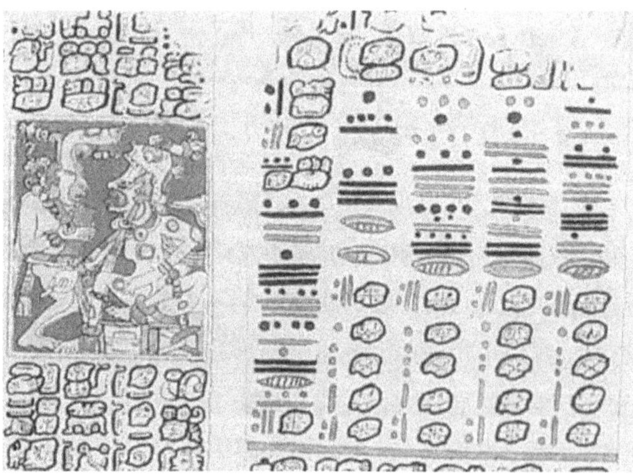

Fig. 1.16. A detail of the Dresden Codex, a Maya document, showing the use of the "shell glyph" as an empty-position sign.

thus entered into contact with their eastern colleagues and learned from them, among other things, the sexagesimal system. They adopted the method for fractions, presumably because they found it more convenient and flexible than the Greek notation for fractions. They used it especially for the measure of angles, thus passing on to future mathematicians the sexagesimal angular units that we still use today. With the system, they also adopted the empty-position sign, which they adapted to their own script. The best documented instance is provided by the major Greek astronomer, Claudius Ptolemy (in the second century CE), who used a little circle surmounted by a bar. This sign probably came from the initial of the Greek work for nothing, *ouden*, and the bar served to distinguish it from the letter *omicron*, which in the Greek system of numerals represented seventy. Yet the sign did not spread outside the highly specialized circle of mathematical astronomers. In the Middle Ages, Arab and European authors had to learn it anew from the Indians. The modern name of the sign bears testimony to its tortuous history. "Zero" is a contraction of the word *zefiro*, the Italian version of the term *zephirum* that Fibonacci said the Arabs used. This word was a transliteration of the Arab *sifr*, which was a translation of the Sanskrit *sunya*, meaning "empty". It is worth noting that *sifr* also gave origin to the term "cipher".

For its early users, up to the seventeenth century, the empty sign was not a numeral. It was a special symbol, which was kept conceptually and graphically apart from the nine digits. The digits were signifiers of numbers, and numbers were entities that, albeit immaterial, were independent of their representations. They referred to numerical quantities, to the "how many" of collections of objects. Unlike them, the zero did not refer to anything outside the system of symbols to which it belonged. It was a place holder, used only in combination with other signs, like a syntactical marker similar to a punctuation sign in a written sentence. Today we are so used to using zero as a self-standing symbol and to thinking of it as one of the numbers that we hardly recognize the difference between an empty-position marker and a numeral. Before the seventeenth century, however, zero could not be considered a numeral for two reasons. Firstly, because if it were taken to be the signifier of something external to the symbolic system in which it belonged, it would have to signify nothing, which was a self-contradictory notion. Secondly, even if one were willing to accept the representation of nothingness, nothingness would not qualify as a number. The prevalent mathematical theory, inherited from the Greeks, formally defined numbers as pluralities of units. Therefore, for the empty sign to become a numeral, change had to occur on two levels. On the level of mathematical notation, symbols had to be granted greater autonomy and power than they would have if they were just passive reflections of pre-existing objects. On the level of mathematical concepts, the concept of number had to be broadened to include more than just collections of units. These two transformations were interdependent, and they occurred in connection with the development of a new branch of mathematics, the branch that we call algebra.

The first reason why zero was not a numeral had to do with the absurdity of representing nothingness. One of the oldest problems of Western philosophy, first raised by the Greek philosophers of the early fifth century BCE, concerned the impossibility of thinking or talking of "what is not". The Greeks and the later Europeans schooled in the Greek tradition were aware that any pretense to represent "no thing" has the effect of transforming an absolute negation into an affirmative. This was a confusion of naming and describing that generated befuddling paradoxes. Since nothing is the negation of everything, it is logical that nothing is not equal to itself (i.e., everything is equal to itself). If nothing is turned into something affirmative, however, that means that nothing is equal to itself. Someone trying to use a written sign for nothing would fall for the same illusion as the one that fooled the Cyclops Polyphemus in the Greek epic poem Odyssey, when he believed Odysseus's cunning fib that "Nobody" was Odysseus's name. No wonder that the algorismists, introducing zero to their readers, studiously avoided to state that it signified nothing. They taught, instead, that it was a sign that did "not signify" by itself, or did "not signify properly", but multiplied by ten the digits to the reader's left. The multiplying function was common to all the digits, but for zero it was used as the defining property. Despite this cautious definition, the danger of nonsense remained close, as is attested by the name of the empty sign in German languages, *null*, which derives from the Latin *nulla*, nothing, and by the English synonym of zero, "naught".

In a purely representational system of numerals, like the Greek and Roman systems, the naught would have been destined to remain philosophically meaningless. In a system that functioned as a technology of calculation like the Hindu-Arabic system, however, the sign had a role to play, and the role provided a meaning. Although not definable by what it signified, zero could be defined by what it did when properly associated with other symbols. On the one hand, it was like a punctuation mark in that it "did not signify", in the sense that it did not refer to anything, material or immaterial, outside the symbolic system to which it belonged. On the other hand, it was more than just a punctuation mark because it participated in the composition of numbers and in arithmetic operations in (almost) the same way as the nine digits did. The definition of unambiguous rules of operation rescued the zero from meaninglessness by providing an operational or symbolic, rather than a representational, meaning. This constituted the first step to equalizing it with the other digits.

Apparently, the Indians had already taken this step. In the mathematical chapters of an astronomical treatise written around 650 CE, the Indian astronomer Brahmagupta gave an *operational* definition of *sunya*, or zero, by listing the rules for carrying out arithmetic operations with it (quoted in O'Connor and Robertson, 2000):

> When zero is added to a number or subtracted from a number, the number remains unchanged; and a number multiplied by zero becomes zero.

He also included zero in his rules to calculate with "fortunes" (or "properties") and "debts", as for example:

> A debt minus zero is a debt.
>
> A fortune minus zero is a fortune.
>
> Zero minus zero is a zero.
>
> A debt subtracted from zero is a fortune.
>
> A fortune subtracted from zero is a debt.
>
> The product of zero multiplied by a debt or fortune is zero.
>
> The product of zero multiplied by zero is zero.

If we interpret Brahmagupta's "fortunes" and "debts" as positive and negative quantities, then most of Brahmagupta's rules are the same rules as we use in arithmetic today, and his operational use of zero is a forerunner of the modern symbolic concept of zero not only as a place-holder but also as a number. At any rate, when the Hindu numerals were transported westward, neither the negative quantities nor the operational definition of the zero traveled with them.

Early Algebra

The concept of number that pervaded medieval mathematics in Islam and in Europe can be illustrated by al-Khwārizmī, the author of the best known *kitab al-hisab al-hindi*, or manual of calculation with the Hindu positional system. Al-Khwārizmī worked at the House of Wisdom in Baghdad, a centre of studies created by the Abbasid caliphs, who vigorously supported the arts and sciences. In an environment that fostered tolerance and learning, and in a cultural climate that encouraged the cross-fertilization of practical and theoretical knowledge, he studied Babylonian and Persian astronomy, as well as Greek, Hebrew, and Sanskrit texts, and built upon the productive encounter of different scientific cultures. Among his many works in mathematics, astronomy, and geography, he authored a famous book, *Kitāb al-mukhtasar fī hisāb al-jabr w-al-muqābala* (*The Compendious Book of Reunion and Reduction*). The Arabic name of the method that he presented in this book, *al-jabr w-al- muqābala* was rendered in Latin as "algebra et almucabola". Shortened to "algebra", it became the name of the branch of modern mathematics taught in our schools as a stepping-stone from elementary geometry and arithmetic toward analytic geometry and calculus.

Historians have different opinions as to when algebra began. Some maintain that it originated four thousand years ago with the Babylonians, whose problem-solving techniques were eventually inherited by the mathematicians of medieval Islam. Others think that it was invented by a Greek named Diophantus in Hellenistic times and re-discovered by Europeans in the Renaissance. Still others believe that it did not get started until it became fully symbolic, in the seventeenth century. This variety of points of view is understandable, since a mathematical discipline is not a static object with rigid boundaries and a definite birth date, but an evolving entity made of many ingredients and interactions. To begin giving order to the historical complexity, and for the purpose of tracing the evolution of numbers, it is useful to suggest that algebra emerged as a distinct area of study, recognized by practitioners with a distinct name, when al-Khwārizmī and others showed an interest not only in applying problem-solving algorithms but also in classifying them and providing Greek-style proofs. This does not mean that the new field emerged from nothing, or that once emerged it remained unchanged over time. It was quite the contrary, as will become apparent from a look at those aspects of the history of algebra that are relevant to the evolution of numbers.

In his *Book of Reunion and Reduction*, al-Khwārizmī explained how to solve problems that asked to find an unknown number, which he called the "root", when an equality involving the root and other numbers was given. Al-Khwārizmī identified three kinds of terms entering the equalities; the square of the root, the root multiplied by a numerical coefficient, and known numbers. He then classified the problems into six standard types, as for example "the square and roots are equal to a number" and "the square and a number are equal to roots", and provided solving algorithms for each type, illustrating the algorithms by means of standard examples that contained only positive numbers. The algorithms were sequences of arithmetic operations on the coefficient and the number, including halving, squaring, subtraction, and extraction of square root, but the examples were chosen so as to involve only positive numbers. Al-Khwārizmī understood subtraction as the "taking away" of a smaller quantity from a larger quantity. The operation of *al-jabr*, which translates into "reunion" or "restauration", consisted of "restoring" the quantities that had been taken away in order to reduce the problem to one of the standard forms, which contained only positive quantities. *Al-mukabala*, "reduction" or "balancing", consisted of eliminating equal terms on the two sides of the equality. The following is an example of a standard problem and its solving algorithm (quoted in Katz 2007, 544):

> Squares and numbers equals to roots; for instance, "a square and twenty-one in numbers are equal to ten roots of the same square." [...] Solution: Halve the number of the roots: the half is five. Multiply this by itself; the product is twenty-five. Subtract from this the twenty-one which are connected with the square; the remainder is four, extract its root; it is two. Subtract this from the half of the roots, which is five; the remainder is three. This is the root of the square which you required, and the square is nine. Or you may add the root to the half of the roots; the sum is seven. This is the root of the square which you sought for, and the square itself is forty-nine.

That all the quantities were understood to be positive is clear also from al-Khwārizmī's proofs. He proved the validity of the algorithms by interpreting each term as a geometrical figure and equating the steps of the algorithm to geometrical "cut-and-paste" modifications that changed the shapes but left the area of the figures unchanged. This cut-and-paste technique came from the legacy of Babylonian practical mathematics, but al-Khwārizmī used it to provide proofs of his algorithms in the style of Greek theoretical geometry. He identified the square with a geometrical square, the root with the side of the square, and the "root" term with a rectangle having the root and the number coefficient as its sides. In the example above, there were two algorithms and two solutions, as illustrated in fig. 1.17.

Since the extraction of square root was identified with finding the side of a square, it produced one positive quantity, and the algorithms were so designed as to lead only to positive solutions. Unlike the Indian mathematicians, al-Khwārizmī and his followers dismissed as impossible the cases in which the subtraction prescribed by the algorithm would imply taking away a larger number from a smaller one. Similarly, they did not regard the result of subtracting a quantity from itself as an object to be operated upon according to rules analogous to the operations with positive numbers.

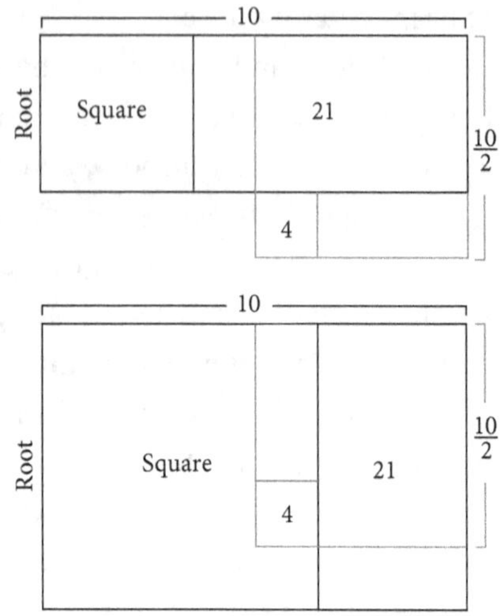

Fig. 1.17. Al-Khwārizmī's geometrical interpretations of the problem, "A square and twenty-one are equal to ten roots" and of the solving algorithms, in the two possible cases: the root is smaller than the half of ten (upper diagram) and the root is larger than the half of ten (lower diagram).

The algebra techniques from Islam were spread by the abacus schools that flourished in Italy in the fourteenth and fifteenth centuries. The teachers of these schools, called "masters of the abacus" or "abacists", used the Italian word "cosa", which means "thing", for the root, and "censo" and "cubo" for the square and the cube. They began writing the problems in shorthand, using the abbreviations *co*, *ce*, and *cu* for these technical terms, and their combination for higher powers, as for example *ce.ce* for the fourth power, or *ce.cu* for the fifth. They also used and $\overline{p}$ and $\overline{m}$, for plus and minus, to indicate addition and subtraction, a long dash for equality, and a cursive "R" with a slash for square root. For example, "The cube and ten things are equal to the square root of four squares" would have appeared more or less as shown in fig. 1.18.

cu.p̄.10.co. —— ℞.4.ce.

Fig. 1.18. An example of shorthand notation in fifteenth century algebra.

As the techniques of the abacists began to be studied in Germany and other European countries, "cosa" became "coss" and algebra became known as "the Cossic Art". The "cossists" began introducing symbols, such as +, –, =, and the "surd" sign, $\sqrt{\ }$. The symbolism was initially diverse and inconsistently used, but gradually it became more uniform and standardized. In the sixteenth century, the abacists and cossists started advancing the techniques beyond the known results; for example, they devised the general methods of solution of problems involving the cube and the fourth power of the unknown. Nevertheless, they continued to accept only positive quantities as numbers.

From a modern point of view, the problems and techniques found in medieval and early modern Arabic and Latin books are easily transformable into algebraic equations. In the example above, we would solve al-Khwārizmī's problem by indicating symbolically the unknown, typically by the letter x, and by "translating" the problem into the quadratic equation, $x^2+21=10x$. We would then transform the specific numerical problem into one particular instance of the general quadratic equation, commonly expressed as $ax^2+bx+c=0$, understanding $a=1, b=-10, c=21$ in the specific case. We interpret the equation as the requirement of finding for which values of the variable x the polynomial ax^2+bx+c assumes the value 0. We can then see that the steps of al-Khwārizmī's algorithm correspond to the operations compactly symbolized by the

quadratic formula, $x=\dfrac{-b+\sqrt{b^2-4ac}}{2a}$.

This exercise in "translation" allows us to highlight the differences between early algebra and modern algebra. The problems of early algebra were expressed verbally; for this reason, early algebra has been characterized as "rhetoric algebra", as opposed to modern "symbolic algebra". The terms involved, in a problem including the unknown, were understood to be fixed and determined quantities, and these quantities were regarded as heterogeneous, for roots, squares, and cubes were given different geometrical interpretations. Thus, the numerical values of the quantities and the operations that could be carried out with them were limited to values and operations that were meaningful according to the geometric interpretation. Different combinations of the terms resulted in different types of questions, which were solved by means of different algorithms. In contrast, a modern equation is a symbolic expression; for this reason modern algebra has been characterized as "symbolic algebra". In it, the terms are homogeneous abstract variables defined only by the symbolic operations that can be carried out with them. This means that different combinations of terms can be regarded as modifications of the same equation, equations can be classified according to the highest power only, and their structure can be studied in a general and abstract way. The greater generality and abstraction of our symbolism is interdependent with a more general and abstract understanding of what a number is. In early algebra the quantities were generally assumed to be integer positive numbers. Depending on the problem, an assortment of other kinds of quantities occurred as results of operations, namely, fractions, negative quantities, square roots of non-square numbers, and square roots of negative quantities. Nevertheless, these objects were either dismissed or dealt with in special ways on the basis of the concrete interpretation of the problem to be solved. In modern algebra all these objects are admitted as "numbers". In ordinary or real algebra, the variables can assume rational and irrational values, positive and negative values, and zero. In complex algebra, imaginary numbers, obtained from the imaginary unit, $i=\sqrt{-1}$, are admitted as well. Thus, we can see that the "translation" from rhetoric to symbolic algebra is not just a matter of expressing the same ideas in another form, with the "form" being just the envelope of some unchanging content. From its medieval to its modern form, algebra underwent a transformation not only in its "language", which became increasingly symbolic, but also in the understanding of numbers that the language embodied. The development of algebraic symbols went hand in hand with the growth of a new idea of numbers that was at the same time more inclusive and more abstract than that of ancient and medieval mathematicians.

Chapter 2

The Evolution of Numbers

2.1 Broadening the Concept of Numbers

The Greek Definition of Number and the Distinction between Number and Magnitude

We have seen in the previous chapter that one of the reasons why medieval Arabic and Latin users of the decimal positional system could not regard zero as a number was the paradox of representing nothingness. The second reason was that they held on to a specific definition of what constitutes a number. In contrast to the use of numbers, which was rooted in everyday practical problem-solving, the definition of numbers was rooted in mathematical theory, specifically the mathematical theory that originated with the Greeks.[1] The Greeks did not just measure and calculate, record quantities and register their rules of operation. They theorized about them as immaterial entities, and wrote philosophically about their nature and about the nature of mathematical knowledge. One of the earliest Greek philosophical schools, the Pythagorean School, had developed a theory of numbers, *arithmetike*, that was founded on an explicit definition of number (*arithmos* in ancient Greek). Following the Pythagoreans, the Greeks defined a number as "a multitude composed of units". This formal description clearly referred to the conceptual entities derived from the practice of counting discrete objects: two, three,

[1]See Chap. 4.

four, and so on. Nevertheless, Greek theoreticians were clear that these conceptual entities were distinct from the counted objects. The philosopher Plato considered them as "forms" or "ideas", that is, ideal entities existing in themselves on a level of reality separate from the world of physical experience. Aristotle, in contrast, argued that numbers were abstractions that the mind derived from the experience of groups of objects.

The Pythagoreans represented numbers as dots arranged in specific patterns, and used the patterns to deduce some of the basic properties of their number theory. The unit itself, the "one" or "monad", was not considered a number but the generator—the building block, so to speak—of numbers. As a theoretical entity, one was considered indivisible. Fractions, although used daily by mathematical practitioners in construction yards, markets, and fields, were not numbers for the theoreticians, but ratios, that is, relations of two numbers. Between two numbers that differed by a unit there existed no other number: numbers were discrete entities.

The Greeks classified quantities into two separate categories, numbers and magnitudes. While numbers were discrete, magnitudes were continuous (fig. 2.1). Practical users of mathematics had always identified magnitudes, as for example lengths and areas, with numbers through the process of measurement. Sometime in the fifth century BCE, however, Greek theoreticians made a discovery that led them to view measurement as inexact and magnitudes as not reducible in general to numbers.[2] They discovered that given any unit of length, no matter how small, there are always lines that are not measurable, that is, that do not contain the unit an exact number of times. They called "commensurable" (i.e., having a common measure) the lines measurable by the chosen unit or by subdivisions of the chosen unit. The relations between commensurable magnitudes, their ratios, are expressible as ratios of numbers. For example, given two commensurable lines, if the respective measures in the chosen unit are three and two, the ratio of the lines is equal to the number ratio 3:2. If the measures are, say, three and one and two tenths, it is possible to change the unit to one tenth of the initial unit and express the ratio as 5:2. Nevertheless, there are also lines that are not commensurable with one another regardless of the unit chosen. The Greeks

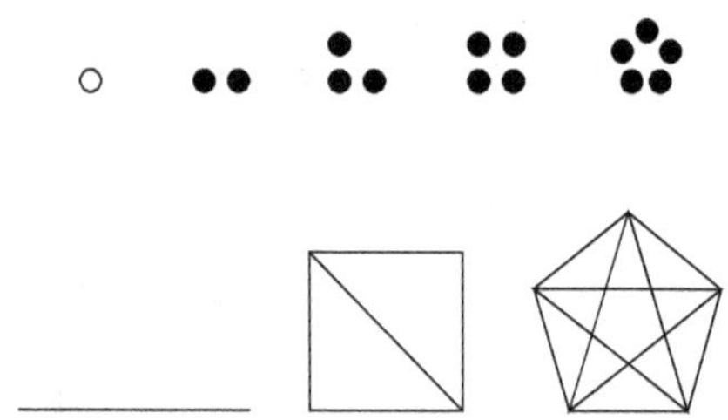

Fig. 2.1. Numbers and magnitudes in Greek mathematical theory.

[2] See Sect. 4.5, "The Discovery of Incommensurability"

called these "incommensurable". Once a unit is chosen, every line will be either commensurable or incommensurable with the unit, and the ratio of an incommensurable line to a commensurable one will not be equal to any number ratio. The best known, and probably the first, instance of incommensurability discovered by the Greeks was that of the side and diagonal of a square. If the unit is chosen so that it measures the side, then it will not measure the diagonal, and vice versa. Another early instance was that of the side and diagonal of a regular pentagon (fig. 2.1). Since numbers were understood as aggregates of units, they were always commensurable with one another. As a consequence of the discovery of incommensurability, the Greeks divorced magnitudes from numbers. Arithmetic and geometry became two separate theories, having respectively numbers and magnitudes as their subject matter. Within geometry, magnitudes were subdivided in two classes, "rational" and "irrational", that is, expressible and inexpressible as number ratios.

The most complete illustration of Greek mathematical theory is provided by the famous treatise *The Elements*, written by Euclid of Alexandria around 300 BCE.[3] Euclid's *Elements* was a systematic compilation of Greek mathematics organized into thirteen "books" (i.e., papyrus scrolls or, in modern terms, chapters) that included geometry as well as the theory of numbers. In the geometrical books, Euclid demonstrated theorems concerning geometrical figures without ever referring to the numerical measure of lengths, areas, and volumes. He expressed the relations between geometric magnitudes of the same kind as ratios and equalities of ratios as proportions, but presented two distinct theories of proportions, one for magnitudes and one for numbers. He devoted Book X to the theory of irrational magnitudes. In it, he classified several kinds of irrational magnitudes and demonstrated, among other theorems, that if the ratio of the squares of two lines is not equal to the ratio of two square numbers, then the lines are incommensurable. Modern authors often write that Euclid demonstrated that the square root of any non-square number is an "irrational number", and, more in general, that the Greeks "discovered" that $\sqrt{2}$, $\sqrt{3}$, $\sqrt{5}$, and so on are "irrational numbers". This, statement conflates modern and ancient mathematics. The concept of irrational number disregards the distinction between number and magnitude that Euclid and other Greeks took great care to keep, and therefore glosses over the long historical process that led to the dismantling of that distinction. Furthermore, for the sake of historical accuracy, we should note that Euclid's notion of irrational magnitudes differed from the modern understanding of irrational numbers for an additional reason, namely,

[3] See Chap. 5.

because he defined as rational not only the magnitudes that were commensurable but also those that had commensurable squares. Thus, the quantities that we represent symbolically as $\sqrt{2}$, $\sqrt{3}$, $\sqrt{5}$ and so on, were rational according to Euclid.

The Decimal Expression of Fractions and Irrational Quantities

The conceptual barrier that the Greeks erected between numbers and magnitudes was demolished by European mathematicians in the seventeenth century. By the time the distinction was openly rejected in theory, however, it had been slowly eroded in practice, first in the Islamic world and then in Europe, by the combined effect of the Hindu-Arabic positional system and the algorithmic style of problem solving in algebra.

Even though the Hindu numeral system was initially introduced for integer numbers only, the mathematicians who used it gradually came to realize that it could be used for fractions as well. For example, al-Uqlīdisī explained in his arithmetic textbook that the positional principle could be exploited to simplify the operations of halving and dividing by ten, with the help of an apex to indicate the division by ten. When halving an odd number, the resulting one half could be written as five preceded by an apex, '5, and when dividing a number by ten the result could be written as the same number preceded by the apex. In the twelfth century, a mathematician named al-Sāmāw'āl ibn Yahyā ibn Yahūda al-Maghribī wrote a *Treatise on Arithmetic* in which he explained that the results of any division could be expressed in decimal form by decomposing the fractional part in tenths, hundredths, thousandths, and so on. He also explained that the division might result in never-ending sequence of digits, as happens for instance when dividing ten by three, but in any case it could be carried on to the desired degree of precision. The understanding of decimal fractions was completed in the fifteenth century by Ghiyāth al-Dīn Jamshīd al-Kāshī, who developed a convenient notation by separating the decimal fractional part from the integer part of a number by means of a vertical bar, just as we do with the decimal point.

Irrational quantities, too, could be represented numerically in decimal form. Algorithms for the extraction of square roots were known since the time of the Old Babylonians; they consisted of sequences of operations that either led to an exact result or could be carried out recursively as many times as needed to produce the desired degree of precision. Al-Sāmāw'āl, for example, calculated the square root of ten up to the sixth decimal place, as three and one tenth and six hundredths and two thousandths and two ten-thousandths and seven hundred-thousandths and seven millionths. He noted, "This method enables us [...] to obtain an infinite number of answers, each one being more precise and closer to the truth than the preceding one." (Quoted in Katz 2009, 270.) Thus, the decimal notation made possible to express irrational quantities in the same form as rational quantities, and this form made it possible to carry out arithmetic operations in a uniform way. Similarly, the algorithmic solution of algebraic problems showed that irrational quantities could operate in the same way as numbers. Al-Khwārizmī, for example, discussed the problem, "I have divided ten in two parts; I have multiplied the one by ten and the other by itself, and the products were the same." (Quoted in Katz 2009, 275.) Since this was a problem of the "square and number equal to roots" type, the appropriate solving algorithm could be applied to it, and it produced as a result the quantity that in modern terms we represent as $15 - \sqrt{125}$ (Al-Khwārizmī discarded the second solution, $15 + \sqrt{125}$, because it could not be a part of ten.) The Egyptian Abū Kamil ibn Aslam (ca. 850–930 CE) studied the different kinds of irrational quantities discussed in Book X of Euclid's *Elements*, but treated them numerically and operationally as if they were numbers. He used for these quantities the Arabic word for "deaf" or "mute"; this word was translated into Latin as "surdus" and eventually made it into English as "surd", which was occasionally confused with "absurd". Although the usage of this term was inconsistent, we may understand it to mean, "irrational quantity resulting from the algorithmic solution of an algebra problem and treated as if it were a number".

The decimal notation allowed mathematicians to disregard in their practice the differences between numbers, fractions, and surds, for it showed that one could operate with these quantities pragmatically as if they were all numbers. In addition, the geometric interpretation of the algebraic algorithms flatly ignored the Greek distinction between numbers and magnitudes. There was an obstacle, however, in that the homogeneity of algebraic operations was not matched by a corresponding homogeneity in the geometric interpretation. Algebraic quantities were interpreted as different kinds of geometrical magnitudes, namely, roots as lines, and squares and products as quadrilaterals. A big step toward finding a theoretical justification of the operational rules was taken around 1000 CE by Abū ʿAbdallāh al-Hasan ibn al-Baghdādī. In his *Treatise on Commensurable and Incommensurable Magnitudes*, al-Baghdādī showed that it was possible to give a homogenous geometrical interpretation to the all the algebraic quantities, roots, squares, and surds. He started by noting the properties of the square roots of the only numbers that were supposed to have square roots, namely, square numbers, such as four, nine, sixteen, and so on. In these special cases, the numbers and their square roots could have a homogeneous geometrical interpretation, for they could be interpreted as commensurable lines. Furthermore, the square roots were the mean proportionals between the square numbers and the unit. Let us recall that the mean proportional is the middle term in a continued proportion. For example, two is the mean proportional between four and one because four is to two as two is to one. In modern terms, 4 : 2 :: 2 : 1. Let us also recall that there existed a theory of proportion for magnitudes as well. So, for instance, it was well known in Euclidean geometry that the mean proportional between two lines segments could be constructed by conjoining the segments, drawing the circle having the conjoined line as a diameter, and drawing the perpendicular line from the common extreme of the two segments to the circle (see fig. 2.2). Thus, proportions could be used as a bridge between numbers and magnitudes, more specifically, as a basis for a common definition of square roots. Using them precisely in this way, al-Baghdādī generalized the relation between square numbers and their roots by defining the square root of any quantity, number or magnitude, as the mean proportional between the quantity and the appropriate unit. This definition ensured that the square root of any quantity would be of the same kind as the quantity. In particular, the square root of a line would be a line, and hence also the square

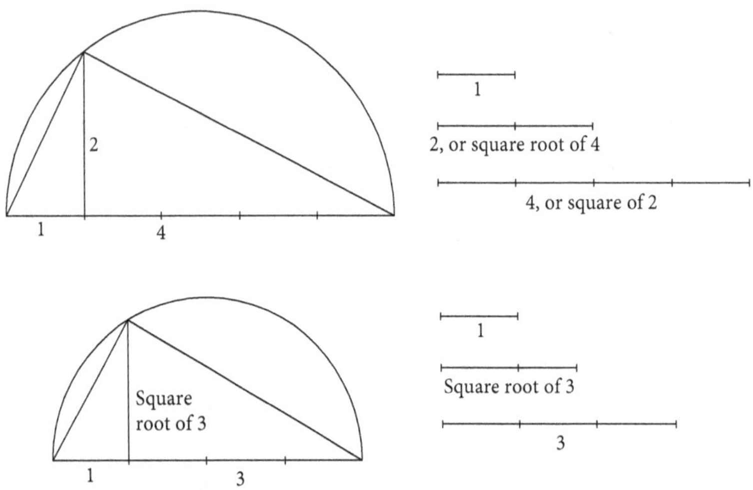

Fig. 2.2. Illustration of al-Baghdādī's definition of square roots as mean proportionals, and the interpretation of rational and irrational quantities as lines, leading to the interpretation of the square roots of non-square numbers (and their square roots, and the square roots of their square roots, etc.) as irrational numbers.

of a line would be a line. Furthermore, since the mean proportional of two lines could always be constructed geometrically, the new definition of square root ensured that any line had a square root. The geometrical interpretation could then be reversed and generalized: the square root of any line could be interpreted as a number (fig. 2.2).

The square roots of lines corresponding to square numbers were commensurable and could be interpreted as rational numbers. The square roots of lines corresponding to non-square numbers were incommensurable, but they could be interpreted as numbers too: they could be a new kind of numbers, irrational numbers. In addition, al-Baghdādī demonstrated that, since the extraction of square root preserved the relation of order between numbers, between any two irrational numbers there were infinite other irrational numbers. The square root of 5, for example, is contained between 2 and 3; therefore, the square root of the square root of 5 is contained between the square roots of 2 and 3. Similarly, the square root of the square root of the square root of 17 is contained between the square root of 2 and the square root of the square root of 5, and so on without limit (fig. 2.3).

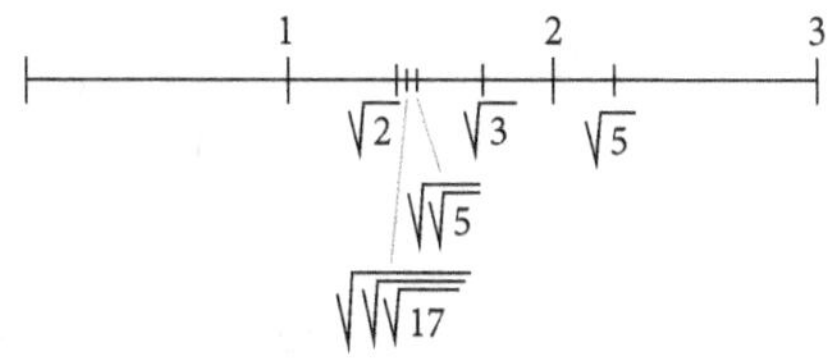

Fig. 2.3. Illustration, in modern symbols, of al-Baghdādī's demonstration that between any two irrational quantities there are infinite other irrational quantities. Let us consider, for example, $\sqrt{2}$ and $\sqrt{3}$. Since $2 < \sqrt{5} < 3$, then $\sqrt{2} < \sqrt{\sqrt{5}} < \sqrt{3}$. And since $16 < 17 < 25$, $\sqrt{2} = \sqrt{\sqrt{\sqrt{16}}}$, and $\sqrt{\sqrt{5}} = \sqrt{\sqrt{\sqrt{25}}}$, then $\sqrt{2} < \sqrt{\sqrt{\sqrt{17}}} < \sqrt{\sqrt{5}}$, etc.

Thus, by providing a homogenous interpretation of all quantities as lines, al-Baghdādī erased the distinction between numbers and magnitudes, and enlarged the concept of number to encompass both rational and irrational quantities.

Stevin's New Definition of Numbers

It is not clear to what extent the European authors were aware of the results of the Islamic mathematicians. The authors of Latin handbooks of arithmetic and algebra, starting with Fibonacci, acknowledged their intellectual debt to their Arabic colleagues, and shared with them a pragmatic attitude in operating algorithmically with different kinds of algebraic quantities, including surds. In the context of real-life money-related problems, they even dealt with negative quantities, which unsurprisingly they interpreted as "losses" or "debts". Al-Kāshī's notation for decimal fractions is found occasionally in European texts from the sixteenth century. At the same time, until well into the sixteenth century, when called to provide a formal answer to the question, "what is a number?" the abacists clung to the Greek definition. A practical handbook published in the Republic of Venice in 1478 and openly addressed to "those who look forward to mercantile pursuits", declared as the basis of "our science" (Swetz 1987, 40–41),

> Number is a multitude brought together or assembled from several units, and always from two at least, as in the case of 2, which is the first and smallest number. Unity is that by virtue of which anything is said to be one.

The advances in solving increasingly complex algebra problems made this definition empty and obsolete. In 1545, the Italian Girolamo Cardano published the general solutions of cubic (third power of the unknown) and quartic (fourth power) problems in his *Ars magna, sive de regulis algebraicis* ("The Great Art, or The Rules of Algebra"). Still, he justified his inability to provide demonstrations beyond the cubic case by the lack of a geometric interpretation for higher powers (Cardano 1968, 9):

> For as *positio* [the first power] refers to a line, *quadratum* [the square] to a surface, and *cubum* [the cube] to a solid body, it would be very foolish for us to go beyond this point. Nature does not permit it.

Furthermore, although Cardano pointed out that "an even power" has "two roots, one plus, the other minus, which are equal to each other," he did not regard the "minus" root as "a true number". As he explained (Cardano 1968, 10),

> But in the case of an odd power, each keep its own nature: it is not a plus unless it derives from a true number, and a cube whose value is minus, or what we call *debitum* [i.e., a debt], cannot be produced by any expansion of a true number. It behooves us to remember this very clearly.

The old definition was openly discarded, at long last, in 1585 by Simon Stevin, a mathematician, military engineer, and financial consultant, who worked for Maurice of Nassau, the Stadtholder of the Dutch Republic. Stevin was put in charge of organizing a new kind of school at the University of Leiden, where students would be taught in Dutch rather than in Latin, and would receive technical training to become engineers, merchants, surveyors, and navigators. As he needed to produce the textbooks for the new school, Stevin had to reconcile the demand of a systematic theory, suitable for teaching, with the pragmatism of the technical professions. He thus reached the same conclusion as al-Sāmāw'āl four centuries earlier, namely that fractions can be written in decimal form, and that arithmetical operations can be carried out with them in the same way as with whole numbers.

Having experienced the power of the positional system, Stevin became convinced that the lack of an adequate system of numerals had stunted the Greek understanding of numbers, and that the time had come to overcome the restrictions of old theory. He believed that a "wise age" had existed in the remote past, when humankind possessed a wondrous science. He wanted to bring back this ancient science, which he believed had been lost during a successive "barbarous age" lasting from the time of ancient Greece to the present. The ancient scientists of the wise age had knowledge of the "0", the sign that gave the positional system its power, but the Greeks had mistaken it for a point, " . ", and had misunderstood the point as the unit from which according to them numbers were generated. To correct this error of the barbarous age, Stevin argued that the real "0" was the arithmetic analogue of the geometric point, and for this reason he did not call it "null" or "nought" or "zero", but "point of number". Just as a line is not changed by the addition of a point, so a number is not changed by the addition of a point of number. Moreover, just as an infinity of points cannot make a line, so an infinity of points of number cannot make a number. Stevin assigned to the point of number the role that the Greeks assigned to the unit, that of "true and natural beginning" of numbers (quoted in Klein 1968, 193).

He then emphatically rejected the old separation of numbers and magnitudes, and defined number as "that by which the quantity of each thing is explained", including both discrete and continuous quantities (quoted in Neal 2002, 35). He argued that the unity was a number, since a part was of the same matter as the whole and the unity was a part of a multitude of units. For the same reason, fractions and square roots were numbers too.

Stevin rejected the notion that numbers were discrete and therefore essentially different from continuous magnitudes. He explained his idea of the relation of number to magnitude by means of an imaginative analogy (quoted in Klein 1968, 195):

> As to a continuous body of water corresponds a continuous wetness, so to a continuous magnitude corresponds a continuous number. Likewise, as the continuous wetness of the entire body of water is subject to the same division and separation as the water, so the continuous number is subject to the continuous division and separation as its magnitude, in such a way that these two quantities cannot be distinguished by continuity and discontinuity.

On the basis of the new definition, Stevin argued that it was incorrect to speak of "surd", "absurd", or "irrational" numbers, for incommensurability did not mean absurdity. Incommensurability was a relationship between numbers, as for example 2 and $\sqrt{8}$, but it did not mean that $\sqrt{8}$ was more "absurd" than 2. To Stevin, a quantity did not need to be expressible as a fully specified numerical value in order to be a number. What mattered was that there existed an operational definition of that value, that is, a definite procedure that allowed the calculation of the value to any degree of approximation. Furthermore, roots, squares, cubes, and so on were not different kinds of numbers, but every number could be regarded as a root, a square, a cube, and so on. In algebra, where the numerical value of a quantity may be indeterminate, quantities may be represented as numbers indicating whether they were to be understood as roots, squares, cubes, and so on, for example ① for a root, ② for a square, ③ for a cube, and so on. Stevin's notation clearly expressed the notion of a geometric progression and anticipated the exponential notation that would soon become common.

2.2 Developing the Algebraic Symbolism

Viète's Vision of Algebra as "The Lost Art of Analysis"

The fifteenth and sixteenth centuries were a time in which many European intellectuals saw themselves as participating in a great renewal of artistic, philosophical, and scientific life. Known as the Renaissance, this cultural movement was fuelled by the rediscovery, translation, and intense study of texts from ancient Rome and Greece, including important mathematical works. Like many of his contemporaries, the Frenchman François Viète was at the same time impressed and dissatisfied with the results of the algebraists. He embarked upon a grand plan to reform algebra, but his goal was a "renovation", a restoring of ancient knowledge that he believed to have been lost. In the preface to the work in which he explained his program, *In artem analyticam isagoge* (Introduction to the Analytical Art, 1591), he wrote (quoted in Klein 1968, 318),

> Behold, the art which I present is new, but in truth so old, so spoiled and defiled by the barbarians, that I consider it necessary, in order to introduce a new form into it, to think out and publish a new vocabulary, having gotten rid of all the pseudo-technical terms, lest it should retain its filth and continue to stink in the old way...

Viète believed that ancient mathematicians had possessed a universal method of discovery in mathematics, the method of "analysis", which they had kept secret and which had then been almost forgotten. Most of what was known of Greek mathematics consisted of theoretical geometry presented according to the method of "synthesis", that is, the procedure of stating general premises (axioms) and deriving from them, as necessary logical consequences, the theorems. Analysis was supposed to be the converse of synthesis. It consisted of preparing a problem for solution by assuming the result as hypothetically known and climbing back the logical ladder, so to speak, up to some already established premises. Viète saw the remnants of the lost "art of analysis", hidden and corrupted, in the techniques of the algebraists ("the barbarians"). Thus, he undertook to study algebra not for solving specific problems that could be useful in practice, but to recover no less than a universal method "to leave no problem unsolved" (Viète, in Klein 1968, 353). Key to the success of this ambitious enterprise was to reformulate algebra in an appropriate "vocabulary", one that could reveal its true power of generality. It was necessary to represent any quantity, be it a number or a magnitude, not only indifferently but also in an indeterminate way, that is, without actually referring to any determinate value but potentially referring to all. To this purpose, Viète introduced for the first time the use of symbols that did not represent specific numbers, as the numerals did, but signified unspecified numbers that were nonetheless assumed to be known. In other terms, not only did he take the practice of the cossists one step further and used single letters to represent the unknowns, but he also pioneered the use of letters also to represent the "knowns", the unspecified numerical coefficients of the equations. He chose to use the vowels, A, E, I, O, U, and Y, for the unknowns and the consonants, B, C, D, etc. for the knowns. Thus Viète would write the equation, $x^3 + cx = d$ as

$$A\,cubus + C\,plano\ in\ A\ aequetus\ D\ solido$$

where *aequetus* stood for equality.

The symbolic representation of indeterminate quantities was a decisive turning point, for it freed the algebraists from the necessity of working with individual numerical examples, and enabled them to focus on the general structure of equations. It opened the way to a new kind of mathematics, no longer concerned either with numbers or with geometric figures, but with symbolic quantities defined through rules of operation irrespective of their nature. But despite his innovative leap, Viète was still confused by multiplication, for which there existed general arithmetical rules but no general geometrical analogue. He still viewed the powers of the unknown as different kinds of quantities, and upheld a "law of homogeneity" according to which only terms of the same kind could be summed or subtracted and compared to one another in equations. Thus, for instance, only cubic terms could be in an equation with a cubic term; for this reason, in the equation above, the coefficient of a linear term had to be specified as a "plane" number, and the constant as a "solid" number.

Descartes

A synthesis of Stevin's view of numbers as continuous quantities and Viète's symbolic representation of indeterminate quantities was achieved by the French mathematician and philosopher René Descartes. Like Stevin and Viète, Descartes approached algebra with the goal of finding a general "art of analysis", a universal mathematical method. His goal was even more ambitious, however, for it was not limited to the traditional domain of the mathematical disciplines. He held mathematical reasoning to be the means to a radical reform of the whole of philosophy. For him, the objects of algebra were neither numbers nor magnitudes, but quantities that could be understood only symbolically, through the definition of the operations that could be performed on them and their results. The objects of arithmetic and geometry, numbers and magnitudes understood in the traditional way, could in some cases provide analogical interpretations of the symbolic quantities and thus aid the imagination. Nevertheless, the quantities could be properly grasped only through what Descartes called "pure intellect". Although he used a symbol for equality that is now in disuse, his equations looked very much like ours. He used Viète's letter symbolism, but started the custom of using the last letters of the alphabet, x, y, and z, as unknowns, and the first letters, a, b, c, etc. as knowns. He adopted the exponential notation for powers, x^3, x^4, etc., except for the square, which he whimsically preferred to write as xx instead of x^2. As we have seen with Stevin, the abandonment of the old manner of representing powers was not just a change of form but reflected a deeper transformation in the understanding of quantities.

Descartes presented his results in *La Géométrie*, which he published in 1637 as an appendix to his philosophical treatise, *Discours de la méthode pour bien conduire sa raison, et chercher la vérité dans les science* ("Discourse on the Method of Rightly Conducting One's Reason, and of Seeking the Truth in the Sciences", or *Discourse on Method* for short). In this work, he showed that any line could be represented by an algebraic quantity, and any algebraic quantity by a line, and that every arithmetic operation had a geometrical analogue. The main difficulty for the correspondence between arithmetic and geometrical operations, as we have seen, was represented by the traditional interpretation of the product of two quantities as not being homogenous with the multiplied quantities. Therefore, Descartes started by redefining the multiplication of two quantities, a and b, as the operation of finding the fourth term of the proportion, $1:a::b:ab$, and providing a corresponding geometric construction in which the analogues of all the four terms were lines (fig. 2.4)

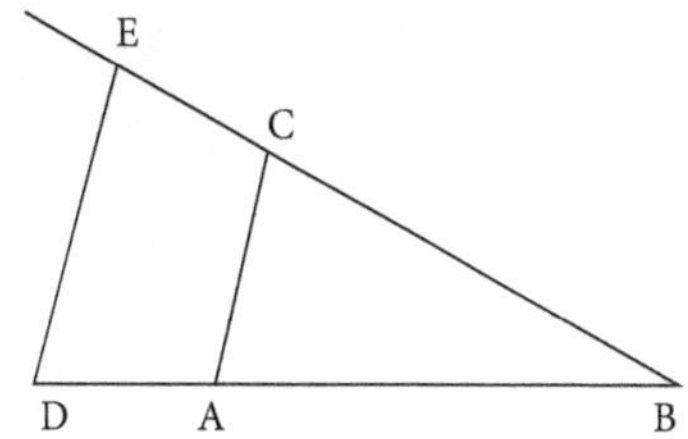

Fig. 2.4. The geometric construction of a product in Descartes's *La Géométrie* (1637).

Representing the unit as the line AB, and the two quantities to be multiplied as BC and BD, the multiplication of BC and BD was carried out by joining the points A and C and drawing the parallel to AC from D. The line BE represented the product because AB, BC, DB, and BE were in proportion thanks to the similarity of triangles ABC and DBE. He further showed how all the terms of the progression $1, x, x^2, x^3, x^4, \ldots$ could be represented and constructed as lines. Thus, Descartes argued that every problem in geometry could be reformulated so that it could be solved by finding the lengths of certain lines. This meant that, thanks to the correspondence between lines and algebraic quantities, every geometric problem could be reduced to an equation. Descartes went on to show that polynomial equations in two variables corresponded to curves on a plane.

With his reformulation of algebra in terms of symbolic quantities and his analogy between symbolic algebra and geometry, Descartes laid the foundations of analytic geometry, the study of curves represented as algebraic equations by means of what we today call a "Cartesian system of coordinates" (*Cartesius* was Descartes's Latinized name). He also launched the study of polynomial equations by means of curves (fig. 2.5).

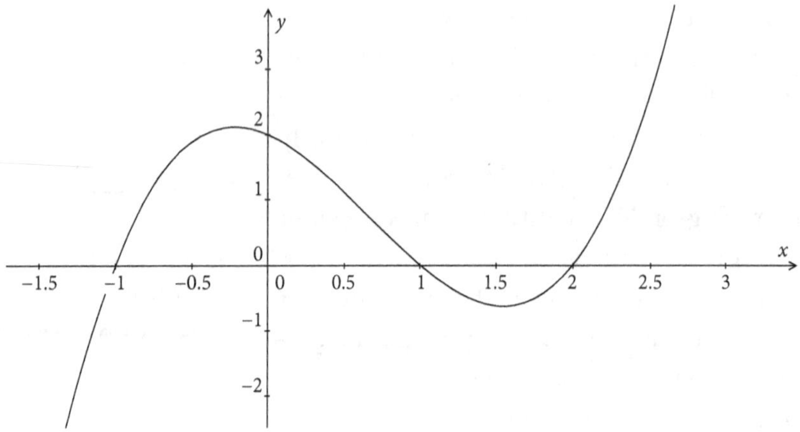

Fig. 2.5. An example of the modern representation of a polynomial equation (in this case, $y = x^2 - 2x^2 - x + 2$) as a curve in a Cartesian system of coordinates. Descartes's original constructions differed from the modern ones because he considered positive coordinates only, and he did not adopt the convention of two perpendicular axes.

Descartes's work became very influential; however, he was not alone in his endeavours. The method of analytic geometry, for example, was formulated approximately at the same time and independently by another Frenchman, Pierre de Fermat, and many aspects of the study of polynomial equations were pioneered by the Englishman Thomas Harriot. The widespread phenomenon of so-called simultaneous discoveries reminds us that mathematics, like the rest of science, is a collective enterprise. Although we may focus on selected historical figures as an expedient to avoid getting lost in the complexity of history, the shaping of new mathematical concepts cannot be reduced to the isolated contributions of a few individuals. This is especially true of the transformation of numbers in early modern Europe, as this period saw the growth of networks of mathematicians and of institutions that promoted the development of the mathematical sciences, the circulation of ideas, and the mingling of the theoretical and practical aspects of science.

Many of Descartes's contemporaries still nominally limited the concept of numbers to positive integers. Descartes himself showed how to find positive and negative roots of polynomials yet still labelled the negative roots as "false". Nonetheless, the willingness to let mathematical objects be defined by arithmetical operations and geometrical constructions made it possible to represent negatives and surds on a number line, along with positive integers and fractions. The Englishman John Wallis, for example, argued that surds as $\sqrt{2}$ as well as anomalous quantities such as the ratio between the circumference and the diameter of a circle—the quantity that we, following a custom that dates from the eighteenth century, represent symbolically as π—are numerically defined through the algorithms that produce their successive approximations or as the sums of infinite series. He also argued that negative quantities could be represented as mirroring extensions of the positive ones by picturing subtraction as a backward motion along the number line.

The analogy between numerical quantities and continuous magnitudes, made possible by the willingness to assimilate indeterminate and symbolical quantities to numbers as long as they obeyed definite rules of operations, laid the groundwork for the development of infinitesimal calculus and of abstract algebra. These two fields converged in the nineteenth century into the formal definition of a new concept of numbers that included the ancient *arithmoi*, the counting numbers, as well as zero, negative numbers, fractions, and irrationals. The new numbers were structured into nested sets, the "natural numbers" as a subset of the "integer numbers", which were a subset of the "rational numbers", which in turn were a subset of the "real numbers"—abstract structures composed of symbolic elements defined by the rules of operation acting between their elements (fig. 2.6).

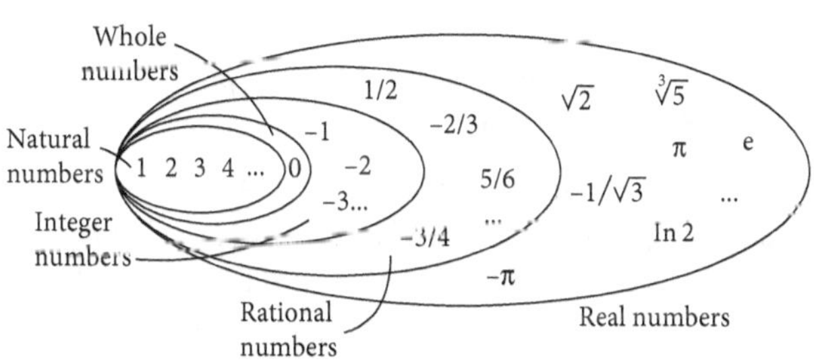

Fig. 2.6. A modern view of the number sets.

Sources

Ascher, Marcia. "Before the Conquest", *Mathematical Magazine* 65 (1992), 211–218. In *Sherlock Holmes in Babylon and Other Tales of Mathematical History*, edited by M. Anderson, V. Katz, and R. Wilson, 98–103. Washington DC: The Mathematical Association of America, 2004.

Burnett, Charles. "The Semantics of Indian Numerals in Arabic, Greek, and Latin", *Journal of Indian Philosophy* 34 (2006), 15–30.

Cardano, Girolamo. *Ars Magna or The Rules of Algebra.* Translated and edited by Richard T. Witmer. New York: Dover, 1968.

Chrisomalis, Stephen. "The Cognitive and Cultural Foundations of Numbers". In *The Oxford Handbook of Mathematics,* edited by E. Robson and J. Stedall, 495–517. Oxford: Oxford University Press, 2009.

Kaplan, Robert. *The Nothing That Is. A Natural History of Zero.* Oxford and New York: Oxford University Press, 2000.

Katz, Victor (editor), Annette Imhausen, Eleanor Robson, Joseph Dauben, Kim Plofker, J. Lennart Berggren (contributors). *The Mathematics of Egypt, Mesopotamia, China, India, and Islam. A Sourcebook.* Princeton and Oxford: Princeton University Press, 2007.

Katz, Victor. *A History of Mathematics. An Introduction.* Third edition. Boston: Addison-Wesley, 2009.

Klein, Jacob. *Greek Mathematical Thought and the Origin of Algebra.* Cambridge, MA and London, England: The MIT Press, 1968.

Kunitzsch, Paul. "The Transmission of Hindu-Arabic Numerals Reconsidered". In *The Enterprise of Science in Islam. New Perspectives,* edited by J. P. Hogendijk and A. I Sabra, 3–20. Cambridge MA: The MIT Press, 2003.

Malet, Antoni. "Renaissance Notions of Number and Magnitude", *Historia Mathematica* 33 (2006), 63–81.

Menninger, Karl. *Number Words and Number Symbols. A Cultural History of Numbers.* Cambridge MA: The MIT Press, 1969.

Neal, Katherine. *From Discrete to Continuous. The Broadening of Number Concepts in Early Modern England.* Dordrecht: Kluwer Academic Publishers, 2002.

Ostashevsky, Eugene. "Crooked Figures: Zero and Hindu-Arabic Notation in Shakespeare's Henry V". In *Arts of Calculation. Quantifying Thought in Early Modern Europe,* edited by D. Glimp and M. R. Warren, 205–228. New York: Palgrave MacMillan, 2004.

Rotman, Brian. *Signifying Nothing. The Semiotics of Zero.* New York: St. Martin's Press, 1987.

Schmandt-Besserat, Denise. "Oneness, Twoness, Threeness. How Ancient Accountants Invented Numbers", *The Sciences* 27 (1987), 44–48. In *From Five Fingers to Infinity,* edited by F. Swetz, 45–51. Chicago: Open Court, 1994.

Swetz, Frank J. *Capitalism and Arithmetic. The New Math of the 15th Century.* La Salle, IL: Open Court, 1987.

Urton, Gary. "Mathematics and Authority: A Case Study in Old and New World Accounting". In *The Oxford Handbook of Mathematics,* edited by E. Robson and J. Stedall. 27–55. Oxford: Oxford University Press, 2009.

Chapter 3

What Do We Know about Egyptian Mathematics?

3.1 The Primary Sources of Egyptian Mathematics

Ancient Egypt, with its pyramids, mummies, and hieroglyphs, has held a special fascination for later cultures. The monumental buildings, funerary furnishings, and other surviving artifacts attest to a high degree of technological competence in architecture, land survey, accounting, metallurgy, pharmacy, and so on. But very few textual sources about these activities have survived, and the scarcity of written information has bred myths and legends about the lost knowledge of this distant civilization. For mathematics in particular, already the ancient Greeks credited the Egyptians with the invention of geometry. The historian Herodotus wrote in the fifth century BCE (Herodotus, *Histories* II, ca. 440 BCE):

> They said that this king divided the land among all Egyptians so as to give each one a quadrangle of equal size and to draw from each his revenues, by imposing a tax to be levied yearly. But everyone from whose part the river tore away anything had to go to him and notify what had happened. He then sent the overseers, who had to measure out by how much the land had become smaller, in order that the owner might pay on what was left, in proportion to the entire tax imposed. In this way, it appears to me, geometry originated.

The term "geometry" means land measurement, and the Egyptians, whose livelihood depended on the annual floods of the Nile, were in a special position to become proficient in the practice of land surveying. A body of state officials had the task of re-establishing the rightful boundaries of agricultural plots every year after the flood. They were called "rope-stretchers" because their characteristic measuring instrument was a rope marked by knots or paint in unit lengths, the cubit and the *khet.* The cubit was the length from middle finger tip to elbow; the "royal cubit" was fixed at about 52.3 cm and was subdivided into 7 palms, while the "small cubit" was about 45 cm, or 6 palms; the *khet* was equal to 100 cubits (fig. 3.1). The work of the surveyors was so characteristic of Egyptian life that the hieroglyphic sign for the number 100 was the pictograph of a coiled rope. The portion of crop that each farmer owned to the state was calculated on the basis of expected yield, which was determined by multiplying the plot extension by a factor that depended on the quality of the soil and other relevant features and conditions. The crop was then measured twice, before and after harvesting, according to standard metrological parameters. These operations were carried out by a team that included the stretcher of the cord, the holder of the cord, the tax scribe, the clerk of the land, and other administrative officials. The stretching of the cord was also a part of the ceremonies that were performed when laying the foundations of official buildings. If long, and hence thick, measuring cords were used for fields and to align walls, probably more precise instruments, such as wooden triangular frames, were used for tasks that required more precision, such as fixing right angles or determining the inclination of pyramids.

Fig. 3.1. A rope-stretcher carrying out a crop survey under the supervision of the scribe Djerserkereseneb. From the scribe's tomb, dated 1400–1390 BCE.

Not only land survey and architecture, but also management of resources and labour, metallurgy, pharmacy, and other activities required mathematical competence. The construction of the Great Pyramid of Giza around the mid-third millennium BCE required several million limestone blocks and must have been a massive logistic enterprise that involved calculations about the relations of base, height, inclination, and amounts of building materials, as well as the number of workers, working hours, food rations and so on. Decorations on tomb walls and statues found in the tombs of high-ranking officials often show figures that can be identified as scribes performing their technical and administrative duties, clearly indicating the existence of a distinct literate and numerate social class trained to serve as a state bureaucracy.

Nonetheless, the perishable nature of their writing media—mostly papyrus, and also wood and pottery sherds—and the humidity typical of the urban areas along the Nile, where literacy and numeracy were practised, have left us with precious few written documents from which to learn about Egyptian mathematics. Besides a handful of small fragments and the archival material of two Old Kingdom temples, we have only two mathematical papyri, the Rhind Papyrus and the Moscow Papyrus. The Rhind Papyrus, so named after the Scottish dealer who bought it in Luxor and sold it to the British Museum in the 1860s, is dated around 1650 BCE but it claims to be a copy of a much older text. The Moscow Papyrus, bought in Thebes by a Russian gentleman in the 1890s and named after the city of the Fine Arts Museum in which it is presently kept, is dated around 1850 BCE. What can be learned from these documents?

Fig. 3.2. The Moscow papyrus.

3.2 The Mathematical Papyri

The Rhind and Moscow Papyri contain mostly mathematical exercises written in hieratic script and arranged in order of difficulty. They were most likely textbooks or notebooks from scribal schools, composed by teachers to instruct students on the methods they would use in their professional life, or by students as learning practice. Here, I will give a short description of the Rhind Papyrus, which is the largest extant mathematical document from ancient Egypt. It opens with a boastful statement by the author, who calls himself "Ahmes" and claims to be copying a century old text: Accurate reckoning. The entrance into the knowledge of all existing things and all obscure secrets." In reality, the content is much more modest: it includes two division tables and 84 problems. The first table lists the results of the division of 2 by all odd integers from 1 to 101, the second the result of the division of 1 to 9 by 10. In this case, as in all other Egyptian documents, only unit fractions are used, with the exception of 2/3. The Egyptians understood and used fractions only as "parts" of whole quantities, that is, "the half of", "the third part of", "the twelfth part of". The so-called "common fractions", 1/2, 2/3, 1/3, and 1/4, had their own specific signs in hieratic script, while all the other fractions were written as the inverse whole number surmounted by a flat oval. Thus, for example, the result of 2 divided by 5 is expressed as the sum of 1/3 and 1/15. Familiar as we are with our kind of fractions, we may think that the Egyptians were making things difficult for themselves by being unnecessarily restrictive. Nevertheless, the Egyptian fractions are arguably more commonsensical than ours, as their magnitude is immediately apparent, and comparisons between fractions is more intuitive in the Egyptian form.

The first group of problems in the Rhind Papyrus is composed of the so-called "*aha* problems", where the term *aha* indicated a numerical quantity to be found under a given condition. One typical example is problem 30: "If the scribe says, what is the quantity of which 2/3 added to 1/10 will make 10, let him hear [the answer]." (Chace 1979, 38.) The answer is then given as the sum of 13 and 1/23, followed by a check. The *aha* problems were clearly intended to teach calculation skills needed to deal with word problems about real-life situations; for example, to find the area of a field or the volume of a granary or to distribute a number of loaves of bread among a number of men. A group of problems is concerned with areas and volumes, and are often accompanied by illustrative sketches. One subset of these deals with pyramids; for example, a translation of problem 56 of the Rhind Papyrus is as follows (adapted from Imhausen 2007, 37):

> *Method of calculating a pyramid*
>
> *360 [cubits] is its base, 250 [cubits] its height. Let me know its inclination.*
>
> *You calculate a half of 360. The result is 180.*
>
> *You divide 180 by 250. The result is 1/2 + 1/5 + 1/50 of a cubit.*
>
> *1 cubit is 7 palms.*
>
> *You multiply with 7.*

We can use these two examples to illustrate some general features of the Egyptian scribal exercises. First of all, it must be noted that the scribes had specific linguistic terms for addition, subtraction, multiplication, division, halving, inverting, squaring and extracting the square root, but they used no mathematical symbols other than the numerals and their parts. The problems and their solutions are always expressed in words and numbers, and the mathematical operations are given in complete sentences. The solutions are given as series of instructions to be followed step by step. Although sometimes a sketch of the figure involved illustrates the procedure, there is no explanation of how the solution is arrived at, and no argument as to why the solution is correct. With the exception of the *aha* problems, all the problems are about particular objects, fields or buildings or quantities of barley, not about abstract geometric figures or abstract quantities. Furthermore, they are always about specific numbers. They are clearly meant to be model problems, or typical examples, used to teach methods that can be applied to other problems of the same kind. In other terms, the exercises are particular problems and they are meant to exemplify a general method. The method is general in the sense that it is understood to be replicable, but is always applied to specific cases. Its generality is not explicitly discussed but only implied by its repeatability. The student is supposed to learn the steps of established procedures, or algorithms, and to become proficient in identifying which algorithm is suitable to every real-life situation and in applying it.

3.3 Scribal Training, Professionalism, and the Practical Character of Egyptian Math

Many aspects of ancient mathematics can be understood by considering the specific historical context in which it was practiced. The professional identity of a scribe was defined by the skills he learned through his special training. These skills distinguished him from the rest of the population and justified his social standing. Since there existed a shared core of mathematical knowledge that was cultivated and transmitted by a self-conscious group of professionals, mathematics in Egypt can be considered a distinct discipline, clearly demarcated from other forms of knowledge and activities. Mathematics was valued in Egyptian society enough to provide support for a class of mathematical professionals. This does not mean, however, that mathematics was valued in the same way as it is today. From the analysis of the mathematical papyri we can infer that the mathematical knowledge imparted by the scribal schools was exclusively of a practical kind; its value resided in its usefulness to the proper administration of the state. This conclusion is confirmed by other written documents, directly tied to practical tasks that required the scribes' mathematical competence, such as the set of four Papyrus rolls, the Reisner Papyri, which contain the architectural specifications and administrative accounting of the construction of a sanctuary.

There also exists, however, an apparently anomalous group of exercises that do not seem to fit well in this picture. Some of the problems in the mathematical papyri mimic the practical problems but are in fact too abstruse and complicated to reflect any realistic situation. They are distinguished from the others by their uselessness. For example, problems 48–50 of the Rhind Papyrus concern the area of a large circular field, a highly improbable shape for a plot of land. This kind of impractical play with mathematical objects is what modern mathematicians identify more readily as "mathematics for its own sake", the kind they hold in highest esteem. Nonetheless, when the position of the scribes in Egyptian society is taken into account, it is possible to give an interpretation that confirms the view of scribal mathematics as eminently practical. Puzzles of this kind—challenging but ludic—are in fact characteristic of specialized professional communities. Some historians argue that the purpose of such useless efforts was precisely in the challenge they posed. The inapplicable but difficult exercises probed the limits of the methods and stretched the skill level required of the scribes. They offered the talented scribe an opportunity to shine beyond mere usefulness. By overshooting the mark, so to speak, these puzzles confirmed the practical skills, reinforced the scribal identity, offered individual distinction and competitive edge within the group of peers. Presumably, they also provided a kind of elitist recreational value, as is shown by problem 79 of the Rhind Papyrus, which reads as follows (Chace 1979, 136):

A house inventory		*houses*	*7*
		cats	*49*
1	*2801*	*mice*	*343*
2	*5602*	*sheaves*	*2301*
4	*11204*	*hekats*	*16807*
total	*19607*	*total*	*19607*

The accepted interpretation of this is that it refers to a mathematical puzzle, which can be written as follows: "*In seven houses there are seven cats. Each cat catches seven mice. Each mouse would have eaten seven sheaves [of grain], and each sheaf, if sown, would have produced seven hekats* [a standard volume measure of grain]. *How many things are these in total?*" Ahmes gives two procedures to arrive at the solution, one on the right-hand column and the other on the left. The first consists of multiplying successively by 7 by 7 etc., and then summing up the results. The other consists of multiplying 7 by 2801, with the multiplication carried out with the method of duplication, that is, as 2801 plus 2 times 2081 plus 2 times 2081. Note that Ahmes made a mistake in the number of sheaves, which should have been 2401. Riddles of this kind are popular among numerically-inclined people in many cultures around the world; one example is the English rhyme,

As I was going to St. Ives
I met a man with seven wives.
Every wife had seven sacks,
Every sack had seven cats,
Every cat had seven kits.
Kits, cats, sacks, and wives,
How many were going to St. Ives?

3.4 Mathematical Form and Content, and the Problem of "Translations"

Traditional history of mathematics takes for granted that the problems of the mathematical papyri, as well as the similar exercises on clay tablets excavated from ancient school sites in Mesopotamia, are best understood if their mathematical "content" is extracted and translated into modern mathematical language. Many popular histories of mathematics follow this tradition, and present Egyptian and Babylonian mathematics by translating the problems and their solutions into the symbolism of modern algebra. The *aha* problem discussed above, for example, can be easily reduced to a modern linear equation in one unknown, $\frac{2}{3}x+\frac{1}{10}x=10$. Thus, these popular histories assert that the ancient Egyptians practised "rhetorical algebra", a primitive and verbal form of algebra, and "were able to solve linear equations."[1] This way of presenting ancient mathematics—quite unlike the respect accorded to the integrity of literary texts from the past—rests on the presupposition that algebra is universal, timeless, and unchanging. Its form can change, but its substance stays the same. In this view, the actual form of old mathematical texts can be disregarded without loss of relevant information, as it is no more than a crude wrapping, and can be replaced with the more suitable modern form.

[1] And no more than linear equations, it is implied, while the Babylonians advanced up to quadratic equations.

A newer current of socially and culturally minded historians regards such "translations" as historical distortions. Historians of this current do not consider the content of mathematics as independent from its form. For them, the verbal expression of the scribal exercises is not just the rudimentary appearance of something unchanging, of which symbolic equations are the true form. They view verbal problems and symbolic equations as different modes of mathematical knowledge, and investigate their interconnections with the respective social and cultural contexts.

The contrast between these two approaches to the history of mathematics is exemplified by the popularity of the already mentioned problems 48-50 of the Rhind Papyrus. To give one specific example, problem 50 reads as follows (Chace 1979, 49. A *khet* is a standard unit of length, a *setat* is a standard unit of area):

> *Example of a round field of diameter 9 khet. What is its area? Take away 1/9 of the diameter, namely 1,*
>
> *the remainder is 8,*
>
> *Multiply 8 times 8. It makes 64.*
>
> *Therefore it contains 64 setat of land*

In this as in all other known examples of this kind of problems, the procedure consists of three steps. A given fraction of the diameter, 1/9, is subtracted from the diameter. The remainder is then squared, that is, multiplied by itself as it is done to calculate the area of a square from the side. The only constant used is 1/9. The modern calculation of the area of a circle consists of applying the formula, $A = \pi r^2$, where A is the area, r the radius, and π is the most famous of all modern mathematical constants. It is certainly possible to "translate" Ahmes's verbal algorithm into a modern-style formula, and then transform the formula by replacing the diameter with the radius multiplied by 2, and compare the resulting numerical constant (the "Egyptian π") with the modern value of π. Such comparison is commonly made, and it is used to support the patronizing claim that the Egyptians were excellent geometers because they made a good, if somewhat awkward, calculation of π. While useful as a didactical exercise about the modern formula, this comparison obscures the character of the Egyptian algorithm. Neither a symbolic formula for the abstract circle, nor anything resembling π appears in the Egyptian text. To paraphrase the historian Annette Imhausen, it is remarkable that so many people are too concerned with projecting their π into the remote past to notice that Ahmes could deal quite effectively with circular areas without using π at all. The next section shows in detail two examples of "translation" of ancient mathematical procedures into modern formulae, namely, the Egyptian problem that we just examined and the Old Babylonian calculations of the area of a circle, and further discusses the limitations of such a ahistorical approach to ancient mathematics.

3.5 Early Calculations of the Area of a Circle and Some Warnings on the "History of π"

In Egypt, Some 3500 Years Ago

Problem 50 of the Rhind Papyrus

> *Method of calculating the area of a circular field of diameter 9 khet*
>
> *You shall subtract 1/9 [of the diameter]*
>
> *It is 1*
>
> *Remainder 8*
>
> *You shall multiply 8 times 8*
>
> *It will be 64*

Translation of the Egyptian "Recipe" into a Geometrical Procedure

Take away 1/9 of the diameter. The remainder is 8 khet. Build a square with side equal to the remainder. The area of the square, 64 square khet, is equal to the area of the circle.

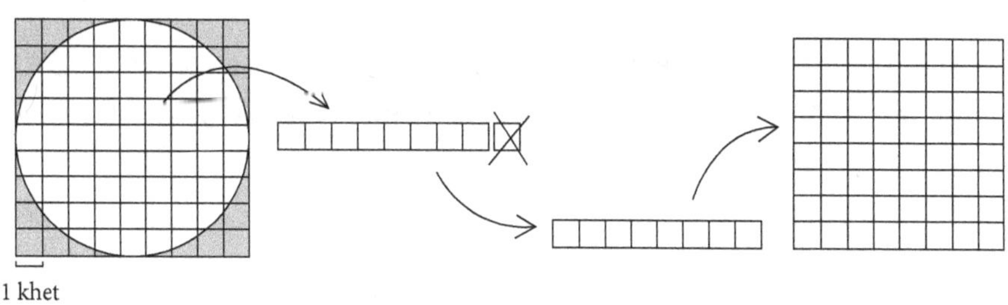

Let us call, in general, the area of a circle A and the diameter d. Let us subtract *1/9* of the diameter from the diameter:

$$d-\frac{1}{9}d=\frac{8}{9}d.$$

Let us square the remainder, and the result will be the area of the circle. Hence, a modern algebraic formula that expresses in general terms the operations of the Egyptian procedure is

$$A=\left(\frac{8}{9}d\right)^2$$

or equivalently,

$$A=\frac{64}{81}d^2$$

Comparison of the "Egyptian Formula" with the Modern Formula for the Area of a Circle

The modern formula for the area of a circle is

$$A=\pi r^2$$

where r represents the radius and π a numerical constant, namely, the ratio between the circumference and the diameter of a circle. Today, we consider this ratio to be a number; we also know that π is an irrational number, which means that it cannot be expressed exactly by a whole number or a ratio of whole numbers. Its expression in decimal form is never-ending:

$$\pi=3.14159265\ldots$$

where the dots indicate an infinite sequence of digits. That is why we use a letter to represent it: its exact value can only be represented symbolically. In any practical calculation, however, a numerical approximation must be used. Commonly used approximations of π are 3.14 and 22/7.

Since the Egyptian formula gives the area in terms of the diameter, to carry out the comparison we have to transform the modern formula so that it also gives the area in terms of diameter. Recalling that $r=\frac{d}{2}$, the modern formula becomes

$$A=\frac{\pi}{4}d^2$$

The Egyptian formula would be equal to the modern one if $\frac{\pi}{4}=\frac{64}{81}$. Therefore, the number that corresponds to π in the Egyptian formula is

$$\pi_{Egyptian}=\frac{256}{81}=3.160493827160494$$

This value differs from the modern value of π by approximately 0.6%. For this reason, modern mathematicians often say—neglecting the differences between the procedure in the Rhind papyrus and the modern formula—that the ancient Egyptians had a pretty good approximation of π.

Meanwhile, in Mesopotamia

The Babylonian Calculation for the Area of a Circle

Some of the mathematical clay tablets from ancient Mesopotamia contain geometrical sketches accompanied by numbers. They have been identified as exercises of scribe students who were learning how to calculate areas. One of the most famous examples is tablet YBC 7302, which displays the sketch of a circle with three cuneiform numbers, the number 3 along the circumference, the number 9 beside the circle, and the number 45 inside it (see fig. 3.3)[2].

Since 45 in the Babylonian sexagesimal notation (which lacked an empty position sign) can be interpreted as 45/60 = 3/4, the numbers can be understood as indicating respectively the length of the circumference, the square of it, and the measure of the area. From this and other similar tablets, we can infer that the Mesopotamian algorithm to calculate a circular area consisted of squaring the circumference and then multiplying the result by a fixed numerical coefficient, namely, 1/12.

Fig. 3.3. Old Babylonian tablet (Yale Babylonian Collection no. 7302, or YBC 7302) containing the calculation of the area of a circle.

Yale Babylonian Collection.

Translation of the Mesopotamian Procedure into a Formula and Comparison with the Modern Formula

If we translate the Mesopotamian procedure into the symbolic language of modern algebra, the "Babylonian formula" turns out to be

$$A = \frac{1}{12}C^2$$

[2]See Chap. 1 for a discussion of Old Babylonian numerals.

where A indicates the area and C the circumference of the circle. For comparison, let us write also the modern formula in terms of C, using the relation $r = \frac{C}{2\pi}$:

$$A = \pi\left(\frac{C}{2\pi}\right)^2 = \frac{1}{4\pi}C^2$$

Therefore, the number that corresponds to π in the Babylonian formula is

$$\pi_{Babylonian} = 3$$

This value differs from the modern value by more than 4%. For this reason modern mathematicians often say that the Babylonian approximation of π is less accurate than the Egyptian one.

Discussion

The comparisons above are exercises in modern mathematics. They do not tell us anything about the ways in which the ancient Egyptian and Mesopotamian scribes thought about geometrical figures and numerical constants. Furhermore, by projecting modern mathematical thinking onto the past, they preclude an open investigation of its historical evolution. If we want to investigate how modern mathematical thinking arose, we need a more historically mindful analysis. Rather than looking for semblances of modern math, we have to place the ancient documents in their historical contexts, paying attention not only to the mathematical content, as seen through modern lenses, but also to the form in which it is presented, the terminology used, the textual and pictorial conventions. We further need to integrate these documents with a study of other relevant historical evidence of the time. For example, the didactical presentation of problem 50 in the Rhind Papyrus shows that the Egyptian scribes did not use any numerical coefficient analogous to our π. They subtracted a piece from the diameter and then built the square of the remainder. When we reduce their procedure to a formula that contains the square of the radius in order to read out a numerical coefficient corresponding to our π, we are looking into the past with tunnel vision. We become blind to a method that was different from our own and yet gave remarkably accurate results.

The Babylonian method, in contrast, appears to be less numerically accurate but it is actually closer to our method, as it did make use of a numerical coefficient that is in a clear sense related to π. This impression may be reinforced by another geometrical tablet of the same period as YBC 7302, which contains the calculation of the circumference from the diameter. In that case, the procedure consists of multiplying the diameter by the numerical coefficient 3, and if we compare that calculation to our definition of π, again we find 3 as the Babylonian value of π. Nonetheless, also in this case, the retrospective comparison is limiting and distorting. There is no evidence that the Babylonian scribes regarded the two numerical coefficients, that of the area calculation and that of the diameter calculation, as expressions of a single constant comparable to our π.

Still, there are differences between the Babylonian way of dealing with circles and ours, and they prove to be at least as interesting as the similarities. For us, the defining characteristic of a circle is the radius, and we calculate area and diameter using π as the only constant. For the Babylonians, in contrast, the defining characteristic, from which other features such as area and diameter were calculated, was the circumference. Furthermore, when we take notice of the form in which the area calculations appears, we note that unlike modern diagrams, the circle in the tablet does not show the center and the radius. These two entities were not involved at all in the area calculations. From these observations and from linguistic analysis of geometric terminology in text tablets, experts of Mesopotamian mathematics have concluded that while modern mathematicians think of a circle as the figure swept by one full rotation of the radius around the centre, Babylonian scribes saw the circle as the figure comprised within the circumference. More in general, they tended to define figures by their boundary lines, such as perimeters and circumferences. As the historian Eleanor Robson put it, whereas modern mathematicians tend to think of geometrical figures "from inside out", the ancient Babylonians thought of them "from outside in".

Finally, the Egyptian and Babylonian calculations are found in documents that were used for exercises in scribal schools. They were part of the institutional training of civil servants, technicians having a precise role in the administration of the state. The purpose of the exercises was that of teaching an algorithm leading to a useful result. In Egyptian and Babylonian sources we find no evidence that the scribes, teachers and pupils, were disturbed by the fact that their procedures produced only approximations but did not lead to exact results. Their results were good enough for all practical purposes. Modern mathematicians' reverence of π derives from the value they place on symbolic abstraction, complete generality, and absolute exactitude. Yet, this value is not universal and unchanging. It is the result of our specific history.

By blindly translating the ancient scribes procedures into our symbolic formulae we erase the fact that they were concerned with effective practical calculations, not looking for any general and absolutely exact formula. It was not their job to find the exact properties of abstract geometrical entities and to prove them with irrefutable logical rigour, as modern mathematicians do. Both the study of abstract geometrical figures and the requirement of mathematical proof only emerged in Greece in the fifth century BCE, as part of a liberal (i.e., non utilitarian) education in a social and cultural context that was very different from that of the ancient Egypt and Babylon.

Sources

Chace, Arnold Buffum. *The Rhind Mathematical Papyrus. Free Translation and Commentary with Selected Photographs, Transcriptions, Transliterations and Literal Translations*. Reston, VA: The National Council of Teachers of Mathematics, 1979.

Imhausen, Annette. "Egyptian Mathematics". In *The Mathematics of Egypt, Mesopotamia, China, India, and Islam: A Sourcebook*, edited by V. Katz, 7–57. Princeton: Princeton University Press, 2007.

Imhausen, Annette. "Traditions and Myths in the Historiography of Egyptian Mathematics". In *The Oxford Handbook of Mathematics*, edited by E. Robson and J. Stedall, 781–800. Oxford: Oxford University Press, 2009.

Ritter, James. "Egyptian Mathematics". In *Mathematics Across Cultures. The History of Non-Western Mathematics*, edited by H. Selin, 115–136. Dordrecht and Boston: Kluwer Academic Publishers, 2000.

Robson, Eleanor. *Mathematics in Ancient Iraq: A Social History*. Princeton: Princeton University Press, 2008.

Rossi, Corinna. "Mixing, Building, and Feeding: Mathematics and Technology in Ancient Egypt". In *The Oxford Handbook of Mathematics*, edited by E. Robson and J. Stedall, 407–428. Oxford University Press, 2009.

Chapter 4

Greek Mathematics

4.1 The Sources of Greek Mathematics

What do we know about ancient Greek mathematics, and how do we know it? On the one hand, we seem to know a lot more about it than about any other mathematical culture because western science is related to it by a continuous tradition of book-learning—a family tree of venerated texts and their accompanying folklore. To mention only the most famous examples, Pythagoras's name is commonly attached to the best known geometrical theorem; the book *The Elements* by Euclid, has been the backbone of mathematical education until the nineteenth century, and its theorems still constitute a major part of elementary math in our schools; and Archimedes was a model to preeminent figures of early modern science, such as Galileo Galilei and Isaac Newton, and is today celebrated as one of the great mathematicians of all time. In mathematics, even more than in other areas of intellectual endeavour, western culture regards itself as the direct descendant of the great Greek authors. On the other hand, very little of the original writings of those authors has actually survived the decline of the Roman Empire and has arrived to us in pristine form. The great library of Alexandria, the almost equally famous one of Pergamon, and all the public and private libraries that existed in imperial Rome were completely destroyed between the second and fifth centuries of our era. What has been handed down is a patchy and often haphazard selection of copies of copies, translations, second-hand accounts, and commentaries that were made in most cases centuries after the deaths of the original authors. The achievements of the Greeks inspired a great deal of admiration in the Islamic world and in Europe during the Middle Ages and the Renaissance. Efforts to interpret, imitate, and then surpass them have unquestionably been a dominant factor in the development of mathematics. Their historical impact, however, has also been exaggerated and distorted. Some overenthusiastic modern commentators have taken the restrictive view that the whole history of mathematics begins, somewhat miraculously, with the Greeks, excluding from consideration anything, Greek or otherwise, that did not fit their preconceptions about what "real" (or "rational", or "scientific") mathematics is supposed to be. In investigating the organic relationship that connects us to those distant ancestors, therefore, we encounter a double challenge: to avoid tunnel vision and to separate legend from fact.

4.2 Practical and Theoretical Mathematics in Ancient Greece

Greek civilization began developing in the early first millennium BCE in and around the Hellenic Peninsula and the islands of the Aegean Sea. It was based on a seafaring society thriving on commerce and spreading its colonies throughout the Mediterranean Sea, especially in Asia Minor (modern-day western Turkey) and Magna Graecia (southern Italy). It was also agitated by political fragmentation and unrest. Historians identify three periods in Greek Antiquity, the Archaic period, from the eighth to the sixth century BCE, the Classical period, from the sixth to the fourth century BCE, and the Hellenistic period, from the beginning of the third century BCE to the fourth century CE. The Classical period was a time of great cultural growth characterized by a flourishing of literature and arts, and by the emergence of Greek philosophy, natural science, and mathematics. The early Hellenistic period saw the maturity of the mathematical developments of the previous two centuries, and is often referred to as the Golden Age of Greek mathematics.

Despite the abundance of historical studies on the subject, many questions about the origins of Greek mathematics are still open. One is, to what extent Greek mathematicians were indebted to the mathematical knowledge of neighbouring civilizations? On the one hand, the Greeks were ready to credit the Egyptians and the Babylonians with teaching them geometry, astronomy, and arithmetic; on the other hand, Greek mathematics displays a great deal of originality, in particular in the abstract, theoretical, and meta-theoretical character of its written works, which differentiates it from the accomplishments of any other culture. Another open question is the relationship between Greek mathematics and Greek philosophy. Yet another open question concerns the relationship between practical and theoretical mathematics, and between theoretical mathematics and the study of the physical world.

The remains of ancient Greek technology testify to a high degree of numeracy and geometrical competence on the part of Greek craftsmen. Archaeological studies reveal, for example, that the Greeks built their colonial cities around the Mediterranean according to detailed grid plans, with the streets intersecting at right angles and forming rectangular city blocks. Greek city planners used a geometrical pattern of straight lines to facilitate orientation and impose order on people's living space. Similar regular patterns can be detected in agricultural land subdivisions, where geometrical order ensured fairness in the distribution of resources. Geometry also enabled feats of engineering, as for example in the construction of an aqueduct on the island of Samos. An engineer named Eupalinos was hired around 540 BCE by the local tyrant, Polycrates, to build a tunnel to bring water to his city from a spring on the other side of a mountain. Two teams excavated the tunnel, which is approximately 1050 m long, working from the opposite ends and meeting in the middle. Eupalinos planned the works by means of extremely accurate measurements and triangulations. Greek architecture is also famous for its buildings, whose harmonious proportions define the ideal of classical beauty. Some historians of architecture maintain that already in the sixth century temples were built so that their width was the mean proportional, or geometric mean, between the length and the height. The front of the Parthenon in Athens, which was constructed in the second half of the fifth century BCE, is inscribed in a "golden rectangle", that is, a rectangle in which the height is the golden section of the base.[1] Greek economy was mainly a commercial economy that made use of money. Therefore, it needed not only clearly defined metrologies for

[1] Let us recall that the geometric mean, or mean proportional, between two quantities, *a* and *b*, is the quantity *c* such that *a* is to *c* as *c* is to *b*. In symbols, $a : c :: c : b$. "Golden section" or "golden mean" is a nineteenth-century name for the part of a segment that is the geometric mean between the whole segment and the remaining part. Fifth century Greek architects applied refinements to basic design proportions to produce optical effects that would enhance the appearance of the constructions. For instance, they inclined the columns inward so that they would look straight when observed from below, and made the corner columns thicker than the others to produce a stronger impression of stability.

the commercial exchange of goods, but also effective calculations for currency. The tasks associated with accounting were performed by specialized professionals, who were hired or owned as slaves by private individuals, public institutions, and kings.

It is clear that in ancient Greek society there existed professionals—surveyors, architects, accountants, and others—who practised mathematical crafts and shared a large body of practical mathematical knowledge. Nonetheless, we know very little of who they were and how their professions were structured. It is likely that they were organized in guilds and trained by apprenticeship. In the case of land surveyors and architects, fragments of papyri have survived from the late Hellenistic period (ca. 100 CE), which apparently come from exercise handbooks similar to those from Egypt and Mesopotamia two thousand years earlier. It is possible that at least some of the mathematical technicians were migrant craftsmen who, starting in the early first millennium, imported into Greece the practical mathematics of the Middle East and continued to transmit their traditional practices to restricted groups of apprentices, generation after generation. Since their teaching was for the most part oral (with the exception of exercise manuals), they have left little trace in literary form.

Numerate craftsmen in Greece did not get their education and did not work within state institutions, as was the case with the Egyptian and Mesopotamian scribes, and they did not enjoy an equivalent social status. Numeracy and literacy were not the exclusive privilege of a special caste. This is testified by works of literature portraying contemporary life, such as theatre plays and tribunal speeches, and by many public inscriptions displaying numbers, such as inventories of temple properties, lists of loans and tributes, records of expenses in public works, and other public accounting documents. A fairly high level of mathematical knowledge was expected of common citizens, who were active participants in the political life of the Greek city states, and who were therefore involved in taxation, public finances, and all the numerical checks and balances needed for the management of the collectivity. Monumental inscriptions containing numerical records of public works indicate that mathematics was experienced by the Greeks as a guarantee of fairness and accountability. Greek literature also bears traces of other related cultural and political meanings of mathematics. It was seen as a means to control the environment and impose order on the world, and as an expression of intelligence and civilization. At the same time, it was perceived as a double-edged sword, which could be put to the service of the common good as well as used to enable trickery and deception.

When the mercantile economy became sufficiently wealthy to support an upper class that practised no useful craft, but read and wrote books for pleasure and enlightenment, the members of the elite developed an ideology of distinction that looked down on the know-how of the productive classes. They theorized that true knowledge—*episteme* in Greek, and later *scientia* in Latin—was the kind of knowledge that is free of the demands of everyday life. Around the mid-sixth century BCE, they began producing a kind of intellectual work that they named "philosophy" (literally, love of knowledge or wisdom). In close connection with philosophy, starting in the mid-fifth century BCE, a subset of Greek thinkers developed a new kind of mathematics that did not aim at the solution of practical problems but was concerned with general truths about abstract objects. This impractical discipline, which we shall call theoretical mathematics, was practised by rare individuals. The historian Reviel Netz made the educated guess that about a thousand theoretical mathematicians were active in the whole Classical and Hellenistic periods. With a few exceptions, they were self-taught and transmitted their knowledge through learned treatises. These were the books that took the lion's share of attention from later admirers of Greek culture. Today, when we talk about Greek mathematics, we mean almost exclusively the theoretical mathematics of the literate tradition. In other words, we adopt the point of view of the two most influential Greek philosophers, Plato and Aristotle, who lived in Athens between the late fifth century and the late fourth century BCE. Practical mathematics became almost invisible to the intellectual tradition of learning initiated by the literate elite.

But Greek mathematics was more diverse than a narrow focus on the theoretical aspects might suggest. Mathematics as we understand it today, as a separate and independent discipline, with its specific institutions and professional categories, did not exist yet. The word *mathema* referred to any discipline of learning, *mathemata* were the things to be learned, and a *mathematikos* was a learned person. Starting in the fifth century BCE, the word *mathematike* acquired a more specialized sense related to numbers and shapes, and some authors defined four *mathemata*, arithmetic, geometry, astronomy, and harmonics, fields of study akin to the mathematical sciences. But the profiles of the *mathemata* did not correspond to those of modern disciplines. Arithmetics was the study of *arithmoi*, a term that we translate as "numbers". The meaning of the term number has expanded over time, however, and the *arithmoi* included only a subset of our numbers, the whole numbers larger than one. An *arithmos*, a number, was defined as a "plurality of units".

One was not regarded as a number but as the origin of numbers, the primary unit from which all numbers were made. Fractions were not considered to be numbers but rather ratios of numbers or specific relationships between numbers. Arithmetic covered the practical arts of counting and calculation, the theoretical study of numbers as abstract entities (the properties of being even and odd, being a multiple or a divisor of another number, decomposition of numbers into sums and products, prime numbers, and so on), and forms of symbolic numerology that today we discount as fantasies. Geometry encompassed the art of measurement, which several Greek authors asserted had been invented by the Egyptian land surveyors, as well as the study of abstract figures with no relation to measurement and of no practical purpose. Astronomy was the study of the heavens; it included the practice of observing and recording the motions of the heavenly bodies, the sun, the moon, the "wandering stars" or planets, and the fixed stars, and the formulation of mathematical models of these motions, two activities that correspond to modern astronomy. Astronomy also had a theoretical and philosophical side, and a practical side. The practical side included timekeeping, navigation, and astrology. Astrology, the practice of forecasting auspicious or inauspicious human events on the basis of the positions of the heavenly bodies, is no longer a scientific discipline; it was dismissed in the early modern age because it lacked any scientific basis. Harmonics was the study of musical harmonies, where the harmonic intervals were associated with simple number ratios. The four *mathemata*, however, did not correspond to academic or professional specialties, nor did they cover all the areas within the reach of mathematical ideas. Although Greek mathematicians were typically quite conscious of the distinction between theory and practice, they were also remarkably versatile. Euclid, for example, wrote books on optics and astronomy in addition to theoretical geometry, and Archimedes is equally famous as an engineer and as a pure mathematician.

Claims of intellectual superiority for theoretical mathematics over practical mathematics were made by the most influential philosophers of the classical age, Plato and Aristotle. Their view, however, did not go uncontested. Whereas Plato portrayed his teacher, Socrates, as a supporter of abstraction, the historian Xenophon reported that Socrates was dismissive of theoretical studies and praised instead the usefulness of geometry for surveying, timekeeping, and navigation. Indeed, Plato's and Aristotle's pleas for the value of theoretical mathematics, as well as the abstract and impractical style that mathematicians adopted when writing theoretical treatises, can be interpreted as defensive strategies against the common opinion that mathematics should serve as a resource in everyday life. The fifth-century BCE playwright, Aristophanes offered sketches of the popular image of a mathematician in his comedies. One of his characters asks a geometry student what geometry is for, and the student replies that it is for measuring land. The character wants to know if it is to allot land to veterans in the colonies, and the student replies that it is for "land in general", evidently alluding to the abstraction and generality of theoretical geometry. The character, however, interprets this as "land for everybody", and comments, "How charming! It's an invention useful for everybody." (Adapted from Cuomo 2001, 19.) We can recognize in Aristophanes's irony a hint of today's squabbles over the social value of a liberal education versus a vocational education. Similarly, in the comedy *The Birds* Aristophanes presents the "air-surveyor", the caricature of a lofty mathematician. The main character, Peisthetaerus, has to build a city in the clouds for the birds, and Meton (the same name as a real mathematician of the time) is one of a number of fancy professionals come to offer their questionably helpful services:

> (Enter METON, with surveying instruments.)
>
> METON: I have come to you...
>
> PEISTHETAERUS: (interrupting) Yet another pest! What have you come to do? What's your plan? What's the purpose of your journey? Why these splendid buskins?
>
> METON: I want to survey the plains of the air for you and to parcel them into lots.
>
> PEISTHETAERUS: In the name of the gods, who are you?
>
> METON: Who am I? Meton, known throughout Greece and at Colonus.
>
> PEISTHETAERUS: What are these things?
>
> METON: Tools for measuring the air. In truth, the spaces in the air have precisely the form of a furnace. With this bent ruler I draw a line from top to bottom; from one of its points I describe a circle with the compass. Do you understand?
>
> PEISTHETAERUS: Not in the least.
>
> METON: With the straight ruler I set to work to square the circle; in its centre will be the market-place, into which all the straight streets will lead,

converging to this centre like a star, which, although only orbicular, sends forth its rays in a straight line from all sides.

PEISTHETAERUS: A regular Thales! Meton...

METON: What do you want with me?

PEISTHETAERUS: I want to give you a proof of my friendship. Use your legs.

METON: Why, what have I to fear?

PEISTHETAERUS: It's the same here as in Sparta. Strangers are driven away, and blows rain down as thick as hail.

METON: Is there sedition in your city?

PEISTHETAERUS: No, certainly not.

METON: What's wrong then?

PEISTHETAERUS: We are agreed to sweep all quacks and impostors far from our borders.

METON: Then I'll be going.

PEISTHETAERUS: I'm afraid it's too late. The thunder growls already.

(He beats him.)

4.3 The Life of the Polis, the Habit of Debate, and the Requirement of Proof

The Greeks were certainly not the first to develop and use sophisticated mathematical techniques. As far as we know, however, they were the first to register meta-mathematical concerns, that is, concerns about the nature of mathematical knowledge and the methodology of mathematical research. As in other scientific fields, such as astronomy and medicine, they raised questions not only about the matters under study (e.g., the properties of even and odd numbers, how to double a cube, etc.) but also about the study itself (what constitutes a number, how mathematical objects are related to the physical world, how general their properties are, and how we know that they are true). Trying to answer these questions, they formulated fundamental distinctions such as those between "magic" and "reason", "empirical" and "rational", "opinion" and "science", and "technique" and "theory". This does not mean that Greek thinkers shed archaic ways of knowing once and forever, or that they were united in following one rational method. The historical record bears evidence of the persistence of magic beliefs (e.g. number mysticism among the Pythagoreans) and of the diversity of approaches (e.g. engineering-style geometry alongside abstract geometry). Nonetheless, Greek science in general and mathematics in particular were characterized by a critical self-consciousness that does not transpire from the sources of previous cultures. The most distinctive achievement of Greek mathematics was the notion of mathematical proof, that is, a logical argument constructed to show explicitely the validity of a proposition. A mathematical assertion that was or needed to be logically proven was called a "theorem", a term which, along with "theory", derived from the verb "to look". Instances of validity checks and demonstrative procedures can be found in the available records of other cultures, but they are informal and not connected to an explicit method. The Greeks were the first to make the requirement of proof systematic and central to mathematical theory.

What historical circumstances fostered the growth of Greek scientific thought, and what factors shaped its character? The relative prosperity of Greek society supported the existence of a wealthy elite, a "leisure class", whose members could devote their time to intellectual pursuit. The use of money in the mercantile economy required a sustained level of numeracy among the general population. Literacy, although restricted to a privileged minority, was sufficiently widespread to encourage the production and circulation of books. The use of a phonetic alphabet has also been singled out as a potential facilitator of analytical and abstract thinking. Commerce and travel contributed to the richness and subtlety of Greek culture because they enabled exchanges with peoples of different customs, social norms, and systems of belief. The combination of these factors certainly created favourable conditions for intellectual life. What distinguished the Greeks from other sufficiently rich and literate societies, however, was the openness of their political systems.

In contrast to the centralized empires of the Near East, the geography of Greece was unfavourable to the formation of large political units. Consequently, in the pre-classical and classical periods, up to the annexation to the Macedonian empire in the fourth century BCE, Greece was divided into a number of small city-states, the *poleis*. The city-states were politically diverse and unstable. Generally speaking, they experienced a variety of forms of government, ranging from tyranny to oligarchy and, famously, democracy.[2] Thus, the citizens were not only exposed to a spectrum of political and legal systems, but actively participated in the formulation and criticism of laws and constitutions. The *agora* (the city square) and the theatre were spaces for public assemblies, discussions, and deliberations. Given the small scale of the polis, a citizen was expected to contribute to public life by voting on matters of domestic and foreign policy, and serving in governing bodies and juries. He was, therefore, experienced in arguing for his own views and assessing arguments between opposing views. The practice of argumentation in front of an audience was so important that professional orators, the sophists, made a living by lecturing on any topic, performing in speech contests, and teaching their skills to the children of the wealthy. From political and legal matters, the habit of questioning and demanding transparency and accountability spread to other intellectual fields, such as medicine, cosmology and, of course, philosophy. Although most debating was done in public and concerned matters of practical consequence, the habit was eventually internalized even by thinkers who disdained common opinions and scorned the opportunistic talk of the sophists. It became a mindset that structured the expression of ideas even of theoretical mathematicians, who devoted their work to non-utilitarian questions and communicated through written mathematical treatises.

We commonly assume that rhetoric, the art of persuasion by means of discourse, is fundamentally different from mathematics. While rhetoric is employed in politics, where interests and emotions are in play, science and mathematics aim at ascertaining hard facts by means of neutral observation and emotionless reasoning. In fact, we often use the adjective "rhetorical" in a pejorative sense, to describe alluring but empty speech. Nonetheless, theoretical mathematics stemmed from the same cultural environment as the art of rhetoric. The self-reflective character of Greek mathematics, with its critical analysis of methods and the insistence on logical proof, can be understood as a response to the habit of debate and the expectation of convincing accounts that pervaded the life of the polis.

We do not know exactly when and how the requirement of proof in mathematics emerged. As already noted, not one fragment of a written source of the early period

[2] We must keep in mind that Greek democracy was restricted to adult male citizens. In other words, it excluded women, slaves, and foreign-born residents.

has survived, and whatever later sources we have are indirect, vague, incomplete, contradictory, and difficult to interpret. One open question, in particular, is whether mathematical proof was an upshot of the requirement of logical demonstrations in philosophy, or whether it was an independent development. The question is complicated by the fact that the earliest extended primary sources we have are the writings of Plato and Aristotle, who dealt with mathematics in the context of their philosophical discussions. Plato, who was Aristotle's teacher, made the first steps in the examination of rational thought, and Aristotle carried the examination forward, producing the first systematic analysis of logic (from the Greek *logos*, which means discourse and reason). The subsequent evolution of an ideal method of theoretical mathematics, the axiomatic deductive method as typified by Euclid's *Elements*, was almost certainly influenced by their teachings.[3] Both Plato and Aristotle, however, took for granted the existence of a substantial body of theorems and proofs, and of a well-developed vocabulary to talk about them.

A long tradition used to attribute the beginnings of Greek mathematics to two legendary figures of the sixth century BCE, Thales of Miletus and Pythagoras of Samos. This familiar narrative enjoys a stubborn vitality in books and websites of science popularization. The following passage from a recent best-seller is a typical example (Roger Penrose, *The Road to Reality*, 2005, quoted in Burnyeat 2007):

> Although mathematical truths of various kinds had been surmised since ancient Egyptian and Babylonian times, it was not until the great Greek philosophers Thales of Miletus (*c.*625–547 BC) and Pythagoras of Samos (*c.*572–497 BC) began to introduce the notion of mathematical proof that the first firm foundation stone of mathematical understanding—and therefore of science itself—was laid. Thales may have been the first to introduce this notion of proof, but it seems to have been the Pythagoreans who first made important use of it to establish things that were not otherwise obvious. Pythagoras also appeared to have a strong vision of the importance of number, and of arithmetical concepts, in governing the actions of the physical world.

The problem with this charming story is that it is made up. Neither Thales nor Pythagoras left any writing, and the fragments of text that have survived from their contemporaries

[3]See Chap. 5 for a discussion of Euclid's Elements.

and close successors contain nothing that can be safely interpreted as being about mathematics.

The long-lasting fame of Thales and Pythagoras as the masters who laid the "foundation stone of mathematical understanding" derives from stories based on other stories, the earliest of which were written several generations after the death of the protagonists. All this storytelling turns out, upon scrutiny, to be mythmaking with very little historical evidence to back it up. As one historian recently noted, the stories about the two grandfathers of western science have the curious property of gaining more biographical "details" as they get more distant from the lifetime of their subjects (Cuomo 2001).

Thales, a political leader in the Ionian city of Miletus in Asia Minor and one of the "seven sages" according to Plato, was credited over time with implausible feats, such as outperforming the Egyptians and measuring the height of pyramids by looking at their shadows, and predicting a solar eclipse. Among his alleged merits was the proving of several geometrical theorems, for example, that two vertical angles are equal and that the two base angles of an isosceles triangle are equal. Yet nothing is known about how he was supposed to have proved them. What we can say for sure is that he gained a reputation as a man of great wisdom. By the time Aristophanes wrote his comedies, his audience knew Thales's name not only as a sage but also as the archetypical mathematician. Pythagoras, the guru of a political-religious "brotherhood" in the colony of Croton in Magna Graecia, in addition to supposedly having superpowers such as being present in two places at once and biting a snake to death, was once believed to have proved the theorem that bears his name—that the sum of the squares of the two short sides of a right-angle triangle is equal to the square of the hypotenuse. The earliest Greek proof of that theorem, however, is to be found in Euclid's *Elements* and Pythagoras had almost certainly nothing to do with it (see next section.)

Most historians today regard the story that places the origins of Greek demonstrative mathematics with Thales and Pythagoras as a myth of origins, or a foundational myth. Foundational myths have social and didactical functions in science. They serve to provide cohesiveness to a field of study, and their narratives are used to arrange in logical order the shared knowledge of the scientific community. This is helpful for didactical purposes, and for the purpose of identifying problem issues and directions of research. Foundational myths are not necessarily fictional. They can contain elements of historical truthfulness, but they should not be taken to be accurate reflections of historical events without independent confirmation. To the best of our current knowledge, if we want a reliable history of the origins of Greek mathematics we need to look at a later time than that of Thales and Pythagoras.

Scholars of Antiquity customarily classify the earliest Greek philosophers into two groups, the pre-Socratics and the rest, according to whether they lived before or after Socrates. The historical information available on the pre-Socratics is very patchy, but it presents no trace, direct or indirect, of mathematical proofs.[4] In contrast, clear evidence of activity in theoretical mathematics begins to surface in the city of Athens around the time of Socrates, the fifth century BCE, and steadily increases from then, reaching a peak at the beginning of the Hellenistic period. In what follows, we will glance at a few developments of the classical period, with the aim of highlighting the distinctive character of Greek theoretical mathematics. First, however, we need to dispel some common misconceptions.

4.4 Pythagoras and the so-called Pythagorean Theorem: Busting the Myth

Pythagoras is a star of scientific folklore. Yet, his historical trail is so muddled by lack of direct evidence, distortions, and imaginative fiction that the reconstruction of a reliable picture is a formidable task. The popular image is that of a fascinatingly duplicitous figure: on one side, the grandfather of hardcore science, demonstrative mathematics, and mathematical laws of nature; on the other, a mysterious purveyor of magical beliefs. Here is how the British mathematician and philosopher Bertrand Russell, following a millenary tradition, introduced him in his *History of Western Philosophy* in the mid-twentieth century (Russell 1946, 38):

> Pythagoras ... was intellectually one of the most important men that ever lived, both when he was wise and when he was unwise. Mathematics, in the sense of demonstrative deductive argument, begins with him, and in him

[4] There are philosophical arguments that are in some sense analogous to mathematical demonstrations. For example, the philosopher Zeno of Elea, a follower of Parmenides, claimed to demonstrate logically that motion is a delusion of the senses by arguing that to traverse any given distance you would first have to traverse half of it, but before you could do that you would have to cover a half of the half, and so on. That, he claimed, would take infinite time. This argument has the logical structure of a *reductio ad absurdum* (reduction to absurdity), which is also one of the earliest forms of mathematical proof. Nevertheless, philosophical demonstrations are not as irrefutable as mathematical proofs. It is in fact possible that mathematical proofs were developed precisely as a kind of argument that would be immune to the endless quarreling that affected philosophical demonstrations.

> is intimately connected with a peculiar form of mysticism. The influence of mathematics on philosophy, partly owing to him, has, ever since his time, been both profound and unfortunate.

This was indeed an inspiring picture. Disappointingly, it has been thoroughly discredited by historical research in the latter half of the twentieth century.

The historical Pythagoras was a charismatic political and religious leader from the Aegean island of Samos, who traveled around the Mediterranean and settled in the Greek colony of Croton in Magna Graecia. There, he founded a sect, the Pythagorean brotherhood, which gained power in several cities of the region and got involved in political violence. That's how we know that Pythagoras and his clan actually existed. It seems fairly certain that Pythagoras believed in metempsychosis, or transmigration of the soul, and that he imparted to his followers a cult-like doctrine. Beyond this, most aspects of the Pythagorean brotherhood remain obscure. They may or may not have been sworn to secrecy. Later authors reported that at some point they divided into two groups, the *akousmatikoi*, the "listeners", forming an inner circle of initiates, and the *mathematikoi*, the "learners", sympathizers of a less spiritual and more philosophical bent, but it is not clear how reliable this report is. It is not even clear to what extent "brotherhood" is the right term, since according to some sources the sect admitted women also. Pythagoras allegedly issued ascetic rules of conduct for the right way of living. Among the many rules that have been attributed to the early Pythagoreans there is some sensible advice, as for instance to make your bed after rising in the morning and not to stir a fire with a knife, as well as some pointlessly persnickety instructions such as to avoid beans and wash the left foot before washing the right, but put on the right shoe first. From this unlikely beginning, somehow, an enduring philosophical school with a strong mathematical bent originated.

Whether following Pythagoras's teaching or not, the early Pythagoreans reportedly took a special interest in numbers, to which they attributed symbolic meanings. By numbers, they meant the pluralities of units that are used for counting: two, three, four, etc. They did not consider one to be a number but the origin of numbers, the unit from which numbers are made. They represented numbers as specks or dots arranged in special patterns, and attached to them symbolic meanings. They divided numbers into even and odd, and considered the even to be female and the odd male. Five, as the union of the first even and odd numbers, two and three, symbolized marriage. Other combinations symbolized justice, reason, and the soul. The number ten, represented by a special symbol called *tetractys* or "mystical tetrad", signified cosmic order and perfection (fig. 4.1).

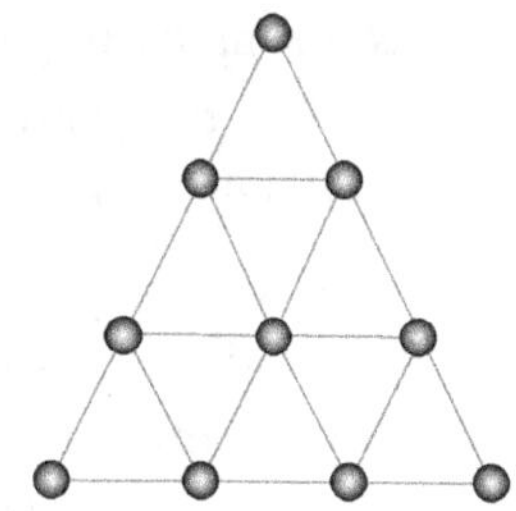

Fig. 4.1. The *tetractys* or "mystical tetrad" of the early Pythagoreans.

The tale that Pythagoras was a mathematician was fabricated by Plato's disciples after the death of their teacher. They sought to validate the mathematical aspects of Platonic philosophy through the myth of a venerable precursor. In particular, they depicted Pythagoras as the originator of the idea that the laws of nature are mathematical and as the pioneer of numerical musical theory. From there, the notion that he discovered or proved the theorem that bears his name followed. What evidence is there for these claims beyond the words of those directly interested in mythologizing Pythagoras?

Aristotle tells us that "the so-called Pythagoreans" believed that "everything is number", that everything was modelled after numbers, and that the properties of numbers were the principles of all things. Although he did not name names, he was probably not referring to Pythagoras but to Philolaus of Croton, who lived one or more generations after Pythagoras and was the first Pythagorean known to have written a book. Philolaus held numerological beliefs similar to those of the early Pythagoreans, for example, he believed that the number seven is or is like (the sense is foggy) the goddess Athena, motherless and childless, because it is not divided nor divides any of the other numbers from two to ten. He also wrote about a numerical cosmology, according to which the world was ordered by numerical ratios in the same way as the musical scale is. He also imagined that everything in the cosmos, including the Earth, revolved around a "central fire". For this reason he was known as an early proponent of the idea that the Earth moves, in contrast to the predominant view (up to the sixteenth century CE) that the Earth is static. Whether Philolaus invented these ideas or adopted them from others, the ideas were most likely a response to those of another philosopher, Parmenides, who lived after Pythagoras. It can be excluded, therefore, that they came from Pythagoras.

Philolaus called the world order "harmonia", clearly referring to the numerical ratios of musical intervals. The poetic idea that the entire cosmos is constructed in such a way as to be structurally similar to the musical scale, or even literally musical—the theory that the rotations of the heavenly bodies, the sun, the moon, and the planets, produce sounds according to the musical scale was later attributed to the Pythagoreans—has influenced many cosmological theories up to the modern age. Most authors agree that it was prompted by one of the earliest and most impressive discoveries of the Greeks, namely, that the main musical intervals, or concords, the octave, the fourth, the fifth correspond to the ratios 2:1, 4:3, and 3:2. This discovery used to be attributed to Pythagoras, of course, and so among his titles was also that of founder of musical theory, harmonics. One old story tells that Pythagoras was passing by the workshop of a blacksmith and heard hammers of different weight striking different notes. He then noticed that two hammers played an octave if one weighed double the other, a fourth if one weighed 4/3 times the other, and so on. Picturesque as it might be, the story is obviously false because hammers do not make sounds proportionally to their weight. Is it possible that the discovery of musical ratios had already been made by Pythagoras's time and that it inspired Pythagoras to give numbers a sacred meaning. Another report attributes the discovery to Hippasus of Metapontum, a *mathematikos* who lived in Magna Graecia one or two generations after Pythagoras. According to the report, Hippasus demonstrated the relation between harmonic intervals and number ratios by means of a more realistic experiment, namely, by making four bronze discs of equal diameter and thickness in the specific ratios, and hanging them to be played as a musical instrument. The only firm evidence we have is that Philolaus knew about the musical ratios, and that he and other *mathematicoi* regarded them as profoundly significant. They associated the musical ratios to the *tetractys*, which they saw as composed by the "union" of 1, 2, 3, and 4.

What about the so-called Pythagorean theorem? Do we owe at least that to Pythagoras? Not really. There is sample evidence that the relation between the sides of right-angled triangles was known and applied in the context of practical geometry long before Pythagoras. Cuneiform tablets containing scribal school exercises from the Old Babylonian period (ca 1800–1600 BCE) show that the geometrical relation between the diagonal and the sides of rectangles was well known in the Middle East at least a thousand years before Pythagoras. To give one example, one of the exercises concerns a pole leaning against a wall in such a way as to form a right-angled triangle. The distance of the lower end of the pole from the wall (the base of the triangle) is found through a solving procedure that consists of subtracting the square of the height at which the upper end touches the wall (the height of the triangle) from the square of the length of the pole (the hypotenuse), and then extracting the square root (fig. 4.2).

Fig. 4.2. Illustration of an Old Babylonian problem involving the calculation of the base of a right-angled triangle from the height and the hypotenuse.

Ancient geometers might have noticed the relation first in its simplest form, that of diagonally divided squares, for example by examining symmetrical tiling patterns such as those that appear on other clay tablets from the same period (fig. 4.3).

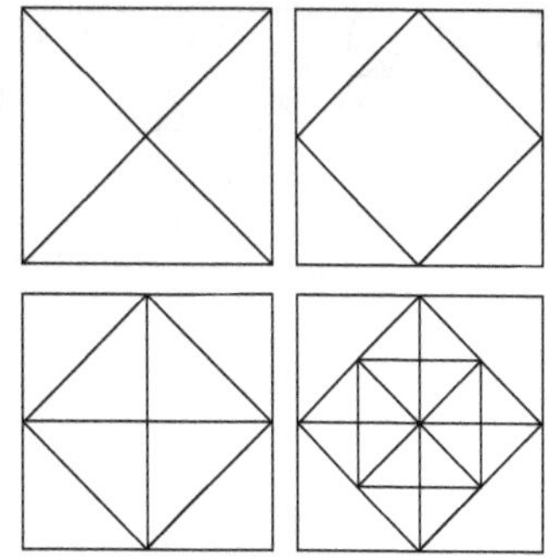

Fig. 4.3. Reproduction of tiling patterns that appear on clay tablets from the Old Babylonian period, ca. 1800–1600 BCE.

(Source: Robson 2008.)

A "block puzzle" or "cut-and-paste" technique can be applied to the picture—decomposing the figure as if it were a physical object composed of triangular pieces, and rearranging the pieces into a new figure—to obtain a convincing, albeit informal, demonstration that the square on the diagonal has the same area as the sum of the squares on the sides (fig. 4.4).

From this special case, using the same technique, the relation can be generalized to diagonally divided rectangles, for example as shown in fig. 4.5.

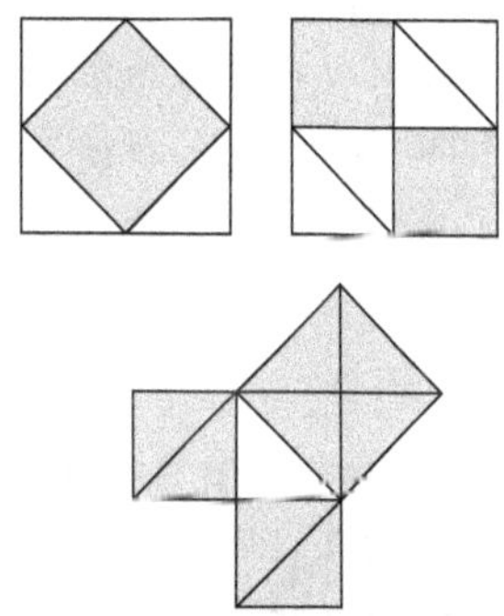

Fig. 4.4. The relation between the square on the diagonal and the squares on the sides of square, i.e., the so-called Pythagorean theorem in the special case of an isosceles right-angled triangle.

This hypothesis is supported by substantial evidence that procedures of this kind, probably first developed by surveyors around 2000 BCE, were part of the institutional training of Old Babylonian scribes. The procedures were used to deal with problems in which the sides of a rectangle were to be found when the area and the sum or difference of the sides were given (quadratic problems).[5] They relied on the trick of decomposing the rectangle and re-composing it into an incomplete square, as shown from left to right in fig. 4.6.

Furthermore, one of the earliest school tablets contains the exercise of finding the sides of a rectangle of which the area and the diagonal are given. The solving procedure consists of cutting the rectangle along the diagonal into two equal triangles, and then showing that the square on the diagonal has the

[5]These problems and their solving algorithm used to be known to twentieth century historians by the anachronistic name of "geometric algebra" because when read through modern lenses they look like quadratic equations. A more historically accurate name is "metric geometry", since when seen in their historical context, the problems belong in the practical geometry of lengths and areas.

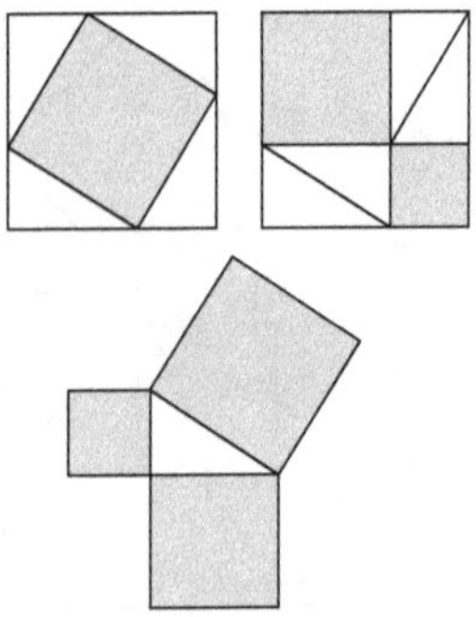

Fig. 4.5. The relation between the square on the diagonal and the squares on the sides of a rectangle, i.e., the so-called Pythagorean theorem in the general case of any right-angled triangle.

same area as the squares on the two sides by means of the block puzzle technique, as illustrated in fig. 4.7.

The Babylonians also had algorithms to calculate the length of the diagonals of squares given the sides (see the discussion of YBC 7289 in the next section.) It seems clear that they had a firm grasp of the general relationship between the square of the diagonal and the squares of the sides.

References to the side-diagonal relation are found in the *Sulbasutras*, the mathematical appendices of the Indian religious text, the *Vedas*, which go back to the first millennium BCE. One version of the *Sulbasutras* contains the statement,

> The rope which is stretched across the diagonal of a square produces an area double the size of the original square.

Another version has the more general statement,

> The rope which is stretched along the length of the diagonal of a rectangle produces an area which the vertical and horizontal sides make together.

An equivalent statement is found in an ancient Chinese mathematical text, the mathematical and astronomical collection titled *The Gnomon of the Zhou*. The dating of this book is very uncertain; it is probably a compilation of texts from different periods, some of which may go back as far as 600 BCE. The text states,

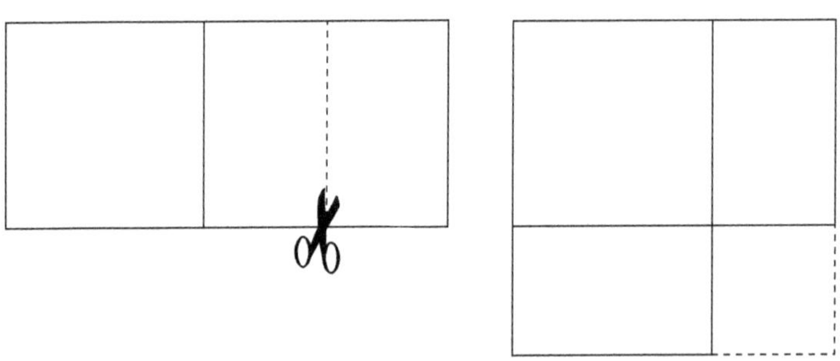

Fig. 4.6. Visual reconstruction of the Old Babylonian *complete the square* procedure for solving quadratic problems.

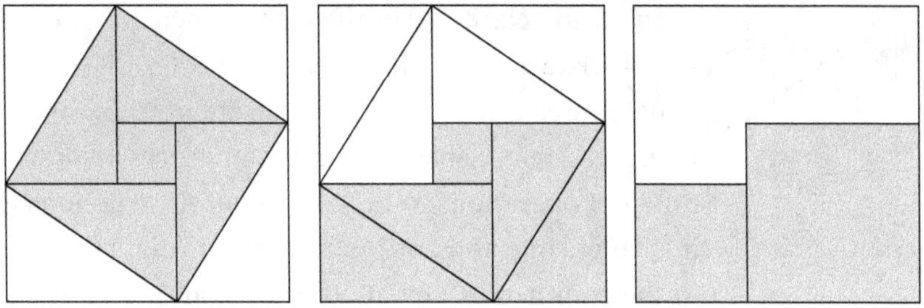

Fig. 4.7. Illustration of the Old Babylonian demonstration that the square on the diagonal of a rectangle has the same area as the sum of the squares on the sides.

> One breaks the rectangle, to take, as base, the width, 3, and take, as height, the length, 4. That which goes through the corners [i..e., the diagonal] is 5.

The text justifies these instructions by showing that the square of the diagonal of the rectangle of sides 3 and 4 has an area equal to 25. Although there is no original diagram in the text, the argument consists of describing a procedure similar to the block puzzle technique, as illustrated from left to right in fig. 4.8:

In other words, the text presupposed a physical or pictorial support that was not a static picture but an object that could be transformed from an initial to a final shape without changing the surface area. The Chinese mathematician Zhao Shuang wrote an extensive commentary on the text in the third century CE. In it, he included the diagram shown in fig. 4.9, adding text to explain how the figure was to be drawn on paper and cut up, how the pieces were to be coloured and then put back together in the final shape:

Referring to the figure, Zhao Shuang stated, "Base and height being each multiplied by itself, summing them up makes the square of the hypotenuse. Extracting the square root hence gives the hypotenuse." (Quoted in Chemla 2005, 142, slightly

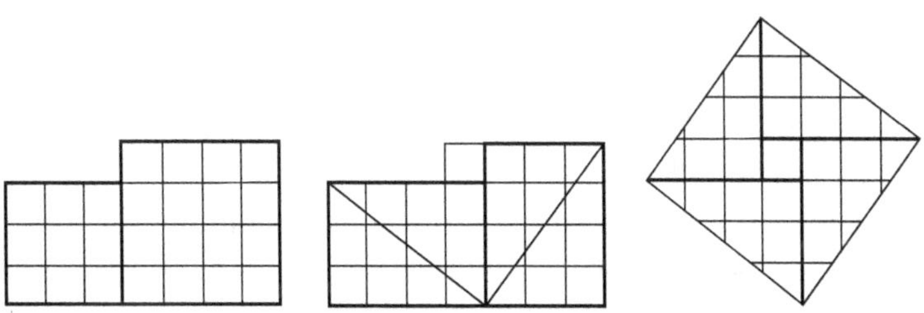

Fig. 4.8. Reconstruction of the procedure described in *The Gnomon of the Zhou* to explain why the diagonal of a rectangle of sides 3 and 4 is equal to 5.

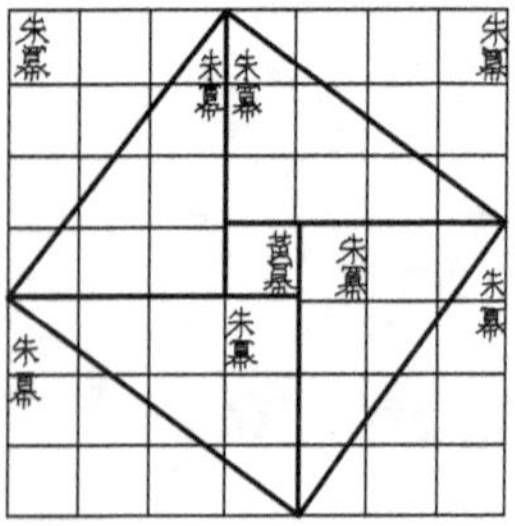

Fig. 4.9. The "Figure of the hypotenuse" from Zhao Shuang's commentary on *The Gnomon of the Zhou* (ca. 200 CE).

adapted for clarity.) He then proceeded to give algorithms to calculate each of the linear quantities, base, height, and hypotenuse, when the other two are given, using the same figure as a general basis for the geometrical demonstration of different operations to be carried out. Instructions for drawing, colouring, and cutting up the same figure are found also in the commentary by another scholar, Liu Hui, on another ancient Chinese mathematical compilation, *The Nine Chapters of the Mathematical Art.* Since it is unlikely that Liu Hui and Zhao Shuang were in contact with each other, it appears that both authors drew from an older, common source.

It seems most likely that the relation we call 'Pythagorean theorem' was discovered independently by many people in different places with the help of some physical or pictorial device similar to those illustrated above. The relation was the basis of algorithms for the calculation of the lengths of the sides of rectangles, and the validity of the algorithms was made evident by means of block puzzle manipulations. We cannot exclude with certainty that Pythagoras was one of the independent discoverers. Nonetheless, we do not have any evidence that he was remotely interested in rectangles, diagonals, or right-angled triangles.

Some modern authors have speculated that Pythagoras' name was linked to the theorem because, although he might not have been the first to discover it, he was the first to prove it. Even this conjecture turns out to be implausible in the light of what we know about the history of Greek mathematics. It does, hoewever, call attention to a distinctive feature of the Greek notion of theorem. In all the early documents about rectangles and triangles, the interest is in the algorithmic procedure for some practical calculation. The description of the algorithm as a sequence of cut-and-paste operations constitutes an implict proof, an underlying argument that derives persuasive force from unstated intuitions based on spatial experience. The Greeks, in contrast, emphasized the proof over the procedure, and developed an abstract and codified style of demonstration that appealed less to sensorial exerience and more to logical relations. Even if Pythagoras did take notice of the side-diagonal relation, it is very unlikely that he would have emphasized the importance of the proof over the algorithm, given the absence of any trace of formal mathematical proof before the fifth century BCE.

The first recorded Greek proof of the relation is found in Euclid's *Elements*, which was written about 300 years after Pythagoras.[6] In *The Elements*, the practical rule was transformed into a theorem. Euclid enunciated it as a general proposition about right-angled triangles (Euclid, *The Elements* I,47):

> In right-angled triangles the square on the side opposite the right angle equals the sum of the squares on the sides containing the right angle.

Then he proceeded to prove it. The proof relies on the main device of Greek demonstrative mathematics, a device that has become familiar to every student of geometry ever since: the lettered diagram (fig. 4.10). The lettered diagram is similar to a block puzzle diagram in that it is a visual aid for building an argument, but the character of both diagram and argument is quite different. The lettered figure is not meant to suggest a physical transformation; it is a static set of carefully labelled elements that are meant to be brought to the attention of the mind sequentially in a rigid order. The relations to be displayed do not rely directly on visual impressions about shapes, line lengths, and equality of areas. They descend from formal definitions, such as the definitions of right angle, triangle, and parallelogram, and formally established properties, such as "all right angles are equal" (a postulate) and "if a parallelogram has the same base and the same height of a triangle, the parallelogram is double the triangle" (a previously proven theorem). While the block puzzle demonstrations rest tacitly but heavily on sensorial impressions, the Euclidean proof is an air-tight chain of logical steps starting with explicitly formulated premises and leading to an inescapable conclusion.[7]

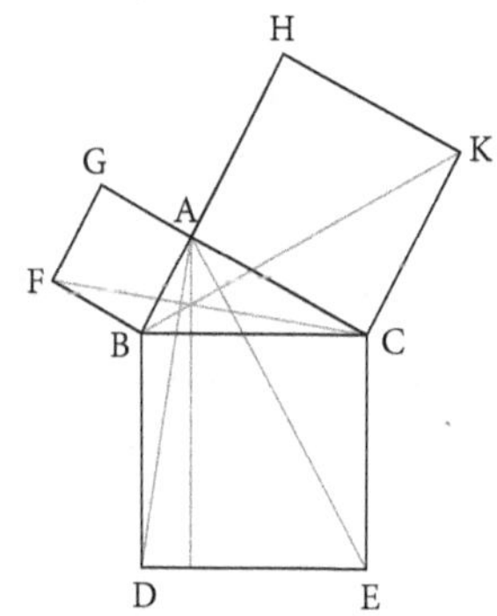

Fig. 4.10. The proof of the so-called Pythagorean theorem in Euclid's *Elements* (ca. 300 BCE).

Finally, while the purpose of the heuristic arguments found in ancient documents was to clarify and support procedures that aimed at useful results, the Euclidean proof presupposed and belonged in a fully developed demonstrative geometry. It appears at the end of the first book of *The Elements* (the last proposition is its converse) after 46 proven propositions, and it appears that Euclid constructed the entire first book, with its logical structure of definitions, axioms, postulates, and propositions, as the apparatus necessary for this proof.

[6] See Chap. 5 for a discussion of Euclid's Elements.

[7] At least, so it is in the prevalent perception of its readers. The alleged completeness and purity of Euclid's logic were criticized in the late-nineteenth and early twentieth-centuries. See Chap. 5 for a brief discussion.

There is no mention of Pythagoras in *The Elements*. The theorem of *Elements* I.47 began to be attributed to our hero only two centuries after Euclid, and it appears that the only basis for the attribution was a couple of verses from an unknown poem of uncertain author and date. The verses narrated that when Pythagoras "found" his theorem, he celebrated by making a "glorious sacrifice of oxen." Besides being hopelessly vague on what theorem was found and how, this story is implausible for various reasons, not least because, as a believer in the transmigration of souls, Pythagoras would probably have objected to the killing of animals for the sake of a theorem.

To conclude, Pythagoras stands out as the classical impersonation of a phenomenon that the sociologist Robert K. Merton called the Matthew Effect, according to which scientific celebrities get credit for the results of more obscure fellows. The expression comes from a passage of the Gospel of St. Matthew, which reads "onto everyone that hath shall be given, and he shall have abundance; but from one that hath not shall be taken away even that which he hath." More specifically, the name "Pythagorean theorem" illustrates one variety of Matthew Effect, Stigler's Law of Eponymy: no scientific discovery is named after its original discoverer.[8] Although dead by now as far as historians are concerned, the myth of Pythagoras lives on in popular imagination.[9] As another British philosopher remarked more recently, "It is hard to let go of Pythagoras. He has meant so much to so many for so long." (Burnyeat 2007, 3.) As teachers have known since the dawn of schools, myths have a role to play in the teaching of math. They put a human face on subjects that students might otherwise find too dry and impersonal. They facilitate the explanation of difficult concepts. They inspire and amuse. Still, they should not be confused with history.

4.5 The Discovery of Incommensurability

Among the results traditionally credited to the early Pythagoreans, the most eventful was the realization that the side and the diagonal of a square cannot have a common unit of measurement. Choosing a unit of measurement is a matter of convention and convenience. When measuring a line, for example, it is always possible to define the unit of

[8] The historian of statistics Stephen M. Stigler pointed out that his law is self-confirmed by the fact that he did not discover it (Stigler 1999).

[9] To be sure, there still are a few historians who support the old view or some revised version of it. The case is intricate even for specialists. To simplify and summarize, since the 1960s the majority view is that there is no evidence that Pythagoras was a mathematician, and from all we know about early Greek mathematics, it is very unlikely that he was. To this, a minority of traditionalists object that absence of evidence is not evidence of absence, and that Pythagoras *could* have worked out at least some of the results that used to be attributed to him.

length so that it is contained exactly a whole number of times in the line. What the early Pythagoreans allegedly discovered is that if the unit is chosen so that it measures exactly the side, it will not measure exactly the diagonal, and vice versa. You may conjecture that choosing a smaller unit would fix the problem. What the early Pythagoreans realized, however, is that no matter how small the unit is, the problem cannot be fixed. Side and diagonal do not have a common measure: they are "incommensurable".

As you may have come to expect by now, we do not know exactly when the discovery was made and how. Some sources point to Hippasus of Metapontum, the early fifth century *mathematikos* who was also credited with demonstrating the harmonic ratios with bronze discs. An old legend tells that Hippasus was drowned at sea, in some versions by his fellow Pythagoreans, in other versions by the gods, either for making the discovery (thus supposedly refuting the alleged Pythagorean dogma that "everything is number") or for divulging it (thus supposedly breaking the alleged vow of secrecy). It is more probable, however, that this dark tale derives from an error of translation.

When two lines are measured by a common unit, the relation of one to the other, that is their ratio, can be expressed as a ratio of whole numbers. For instance, if a line is 6 units long and another is 4 units long, the ratio of the first to the second is equal to $6:4$. Choosing the largest common unit, it is possible to express the ratio as the minimum number ratio, which in this case is 3 : 2. Since side and diagonal of a square are incommensurable, their ratio cannot be expressed as a ratio of (whole) numbers. The Greeks called the ratio of two incommensurable quantities *alogos*, "unaccountable", or *arhetos* ("inexpressible"). We call it "irrational"; conversely, we call quantities that are not irrational "rational". In their mathematical sense, these words have nothing to do with reason or lack thereof; they only refer to whether or not the relationship of one quantity to another can be expressed as a ratio of whole numbers. More precisely, in mathematical usage an irrational quantity is a quantity that is incommensurable with the unit. The discovery of incommensurability consists of the realization that certain lengths, as for example the diagonal of a unit square (the square having the side equal to one unit), are irrational quantities. For this reason it is also often described as the discovery of "irrationality". Mathematicians of the early twentieth century speculated that the discovery of irrationality caused a crisis in Greek mathematics because it shattered the Pythagorean doctrine that everything is number. There is no historical evidence of this alleged crisis, however. On the contrary, Plato used the discovery of irrationality (in the mathematical sense) as an example of what can be achieved by rational thinking (in the more general sense), and as something every educated person should know about. Plato was very impressed by the power of mathematics; he gave it an important role in the education of the ruling class, and promoted it in the school that he founded in Athens, the Academy. Indeed, the discovery of irrationality gave rise to fruitful areas of research in his school. One of his pupils, Theodorus of Cyrene, proved that there are other irrational quantities in addition to the ratio of side and diagonal of a square. He showed that the sides of all the squares, the areas of which are non square numbers up to 17, are irrational. Theaetetus, another pupil of the Academy, produced a theory of irrational quantities, which was later reported by Euclid in Book X of *The Elements*. Numbers, in the Greek sense of the term, cannot be irrational because they always have the unit as a common measure. Hence, the Greeks divided quantities into two classes, discrete quantities or "numbers", and continuous quantities or "magnitudes". They further classified magnitudes into rational and irrational, and irrational quantities into several subclasses. In the time that separates us from the Greeks, the concept of number has greatly expanded, and it has merged with continuous magnitudes again.[10] Today, we classify numbers into increasingly inclusive sets, natural, integer, rational, irrational, real, and more. We represent the real numbers as points on a continuous line. Our concept of number has many ingredients, a large part of which grew from the ramifications of the ancient Pythagorean discovery of irrational quantities. The long history that connects us to that event is rich in conceptual articulations; yet, in popularizations of mathematics, it gets frequently flattened into the oversimplifying assertion that "the Pythagoreans discovered irrational numbers."

To return to the question of how incommensurability was discovered, a hint is provided by Aristotle, who wrote that mathematicians were able to prove that the diagonal of the square is incommensurable with the side by showing that "if it is assumed to be commensurable, odd numbers will be equal to even." (Aristotle, quoted in Cuomo 2001, 33.) Aristotle's brief allusion points to a kind of proof known by the Latin expression, *reductio ad absurdum* ("reduction to absurdity"), or as indirect proof or proof by contradiction: if the consequences of denying a proposition are proved to be absurd or contradictory, then the proposition is proved. The proof to which Aristotle referred might have been a special case of a more general proof, which is given by Euclid in Book X of *The Elements* (Proposition X, 9). It is indeed possible that it was the original proof, since it requires only the so-called Pythagorean theorem, the properties of ratios, and the properties of even and odd numbers, which were all specialties of the early Pythagoreans. The idea of a *reductio ad absurdum* is to deny the proposition

[10]See Chap. 2 for a discussion of the evolution of numbers.

and to show that the denial leads to an impossibility. In this case, the proposition is that side and diagonal of a square are incommensurable; its negation is that side and diagonal have a common measure. If that were the case, then the ratio of diagonal and side would be expressible as a ratio of two (positive whole) numbers. In this hypothetical case, it would be possible to assume without loss of generality that the largest common unit of measure has been chosen, so that the ratio of the two numbers would be the lowest possible hypothetical number ratio representing the diagonal-side ratio. This means that the two numbers could not be both even numbers. It would also be true, however, that the ratio of the squares of these numbers would be equal to the ratio of the squares of diagonal and side. The Pythagorean theorem tells us that this ratio is 2 (fig. 4.11). This means that the two numbers would have to be both even.[11] Therefore, the hypothesis that side and diagonal are commensurable leads to the impossible conclusion that one of the two numbers has to be both odd and even. Hence the hypothesis is false.

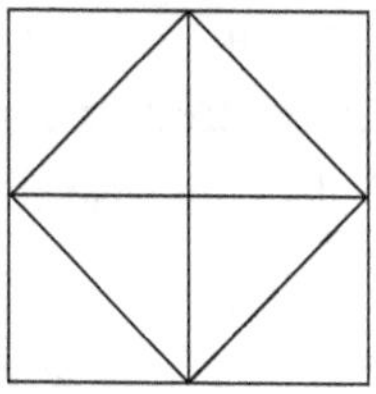

Fig. 4.11. The side and diagonal of a square are incommensurable. This follows from the fact that the ratio of the squares of the diagonal and the side of a square is equal to 2.

For us, the length of the diagonal of a unit square is an irrational number. We represent it with the symbol $\sqrt{2}$. For the Greeks, irrational magnitudes were quantities that could not be represented in numbers. The discovery of irrationality was a turning point in the development of Greek mathematics because it drastically separated theoretical geometry from arithmetic. From then on, Greek geometers produced a great deal of theoretical results concerning ideal figures, lines, angles, polygons, circles, and so on, without ever referring to the numerical expressions of lengths and areas. They compared figures of the same kind, but studiously avoided replacing the figures with numbers expressing their measurements in any given unit; they respected the distinction between magnitudes and numbers. They expressed the relations of magnitudes of the same kind as ratios and the equality of ratios as proportions, remaining aware that only in some cases the ratio of two magnitudes is equal to a number ratio. Thus, the discovery of incommensurability was a major turning point that marked the departure of Greek theoretical mathematics from practical mathematics.

[11] To see why this is, let us call the two numbers n and m. The ratio of diagonal to side would then be represented by $n : m$, and the Pythagorean relation between the square of the diagonal and the square of the side would be $n^2 = 2m^2$. This means that n^2 is an even number. Since only the square of an even number can be an even number, n has to be an even number. Moreover, n^2, being the square of an even number, has to be a multiple of 4. This implies that m^2 is an even number, and hence that m is an even number too.

Fig. 4.12. The Old Babylonian tablet YBC 7289, ca. 1800–1600 BCE.
Yale Babylonian Collection.

Over a thousand years before the Greeks discovered incommensurability, the Old Babylonian scribes knew how to calculate the length of the diagonal of a square. A famous tablet from a scribal school, now preserved in the Yale Babylonian Collection (YBC 7289), shows a square, its diagonal, and sexagesimal numbers along the side and the diagonal (fig. 4.12). The content of this striking tablet was in all probability a school exercise. The student calculated the diagonal of the square given the side, using a typical Babylonian method of multiplying the side by a given coefficient, which he would have looked up in a coefficient table. In this case, as the side was 30, the diagonal turned out to be 42.43 (42; 25 35 in sexagesimal notation). The coefficient used is displayed along the diagonal: 1.41422 (1; 24 51 10). This number corresponds to our $\sqrt{2}$. More precisely, it is an approximation of it, remarkably correct to the fourth decimal digit. As YBC 7289 together with other school tablets testify, the scribes mastered a recursive algorithm capable of producing the sides of a square of given area, that is of extracting the square root of any number, with any degree of approximation. Had the scribes wanted a better approximation, they could have iterated the algorithm to further reduce the difference between 2 and the square of the coefficient. They must have judged, quite rightly, that the calculated value was good enough for any practical purpose. There is no evidence that they wondered whether the calculation could ever be carried far enough to produce the exact length of the diagonal rather than just an approximation of it. As far as we know, worries of this kind did not perturb the practical world of scribal mathematics. For the Greek theorists, however, questions of utterly impractical exactitude took centre stage. Having learned from them, today we know that irrational quantities can neither be expressed exactly as ratios of whole numbers, nor as a finite sequence of digits in a positional system of numerals.

4.6 Hippocrates of Chios, the First *Elements of Geometry*, and the Squaring of the Circle

The earliest mathematical proofs of which we have credible information are from the work of two mathematicians who lived between the fifth and fourth centuries BCE. Their names are Hippocrates of Chios and Archytas of Tarentum, and their results concerned the first two of three famous problems which are known as the classical problems of Greek mathematics, the squaring of the circle (you will recall that Aristophanes's air-surveyor boasted that he was going to "square the circle" with his ruler), the "duplication of the cube" (more about this below), and the "trisection of an angle" (which we shall omit for the time being.)

Hippocrates of Chios (not to be confused with the famous physician, Hippocrates of Cos) proved that circles are to one another as the squares of their diameters. This result was a crucial step toward the solution of the most famous problems of classical Greek mathematics, the problem of squaring the circle or of the quadrature of the circle, that is, of finding the area of the circle (fig.4. 13).

Hippocrates did not arrive at the full solution of the problem. He was, however, able to "square lunes", that is, to find the areas of circular figures that are shaped like crescent moons. In addition to these remarkable feats, he was the author of a collection of geometrical theorems titled *The Elements of Geometry*, the first of a genre of geometrical literature of which Euclid wrote the most famous example. The "elements" were those theorems that were used to prove other theorems. Hippocrates must have felt the need to bring order to what was then an exciting new field, the field of geometrical theorems or demonstrative geometry, which had grown quickly but unsystematically during his lifetime.

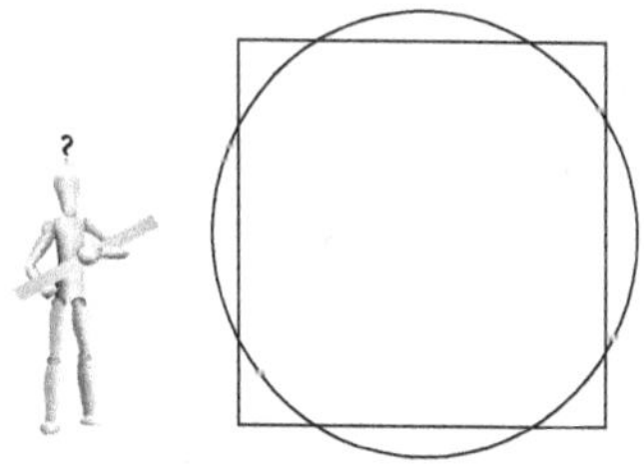

Fig. 4.13. One of the classical problems of Greek mathematics: squaring the circle, i.e., finding the area of the circle. Hippocrates of Chios (ca. 470–410 BCE) proved that circles are to one another as the squares of their diameters.

4.7 Archytas of Tarentum and the Duplication of the Cube

The second historical example of mathematical proof was the solution of another classical problem, the duplication of the cube, by the Pythagorean Archytas of Tarentum. According to legend, the problem originated from the answer of an oracle to the citizens of the island of Delos. Delos was suffering from the plague, and when the citizens asked the oracle what they had to do to stop the epidemics, the oracle replied that they had to double the size of an altar to Apollo, which was of cubic shape. The problem, then, was one of solid geometry: what is the length of the side of a cube of volume twice as large as a given cube (fig. 4.14)?

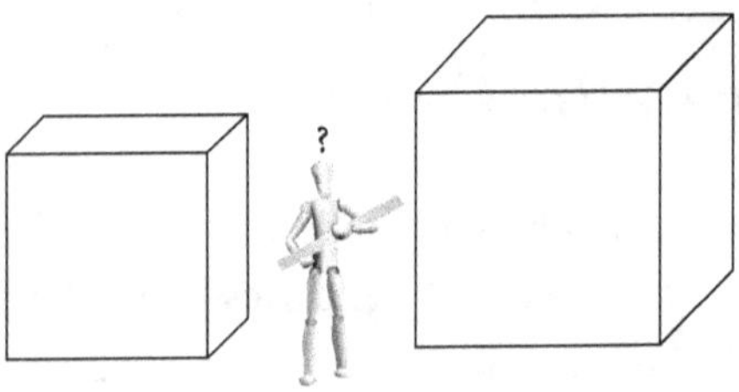

Fig. 4.14. Another classical problem of Greek mathematics: doubling the cube.

Hippocrates made the first step. He reduced the problem to that of finding two mean proportionals between the side of the given cube and its double. To clarify what this means, let us consider first the simpler problem of doubling a square. If a is the side of the original square, the side of the square having an area twice as large will be the mean proportional, or geometrical mean, between a and $2a$, that is, the length b such that $a:b::b:2a$. It can be seen that this is the case by using the property of proportions, namely, that the product of the means (the second and third terms) is equal to the product of the extremes (the first and fourth term). Extending the problem to three dimensions requires two mean proportionals, b and c. If a is the side of the original cube, the side of the double cube will be a length b such that $a:b::b:c::c:2a$. In this case, in fact, the cube of b will be equal to two times the cube of a, as can be seen using the properties of proportions. Therefore, the challenge becomes that of finding the two lengths, b and c, which are the intermediate terms of a continuous proportion between a and $2a$.

Archytas's solution was the first and the most admired. It employed an imposing geometrical construction in three dimensions, consisting of two rotating figures, a triangle and a semicircle, and their intersection with a fixed semi-cylinder. (We do not need to get involved with the details of this daunting construction. See a simplified version in fig. 4.15). What is remarkable is that the results of Hippocrates and Archytas, despite being the earliest documented instances of deductive mathematics, demonstrated already an impressive mastery of the method of proof.

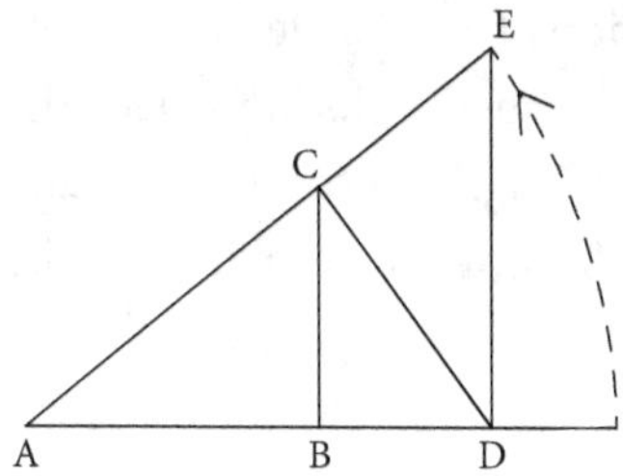

Fig. 4.15. A simplified version of Archytas's method to duplicate the cube by finding the two mean proportionals between the side of the original cube, AB, and the line twice as long, AE. By letting AE pivot around A, it is possible to find the two points C and D at which CB and ED are perpendicular to AD and CD is perpendicular to AE. Then, thanks to the similarity of triangles ABC, ACD, and ADE, AB:AC::AC:AD::AD:AE. The side of the cube of double volume will be AC.

In late Antiquity, the story began to circulate that Plato criticized Archytas's solution for being "mechanical". What this was supposed to mean is not clear, as "mechanical" might be interpreted as "requiring mechanical implements" or as "involving movement". In any case, the story was used as historical underpinning to a long and ultimately unsuccessful quest to solve the problems of squaring the circle and duplicating the cube solely through the application of Euclid's definitions, axioms, and theorems in a finite number of steps. This restrictive prescription—known as the "straightedge and compass method", for Euclidean constructions can be carried out with the use of these two idealized instruments—was the product of an excessive devotion to Euclidean mathematics. The mathematicians of the Classical and Hellenistic periods do not seem to have known of any restriction of this kind, for they produced several inventive solutions availing themselves freely of both motion and instruments.

4.8 Plato and Mathematics

Plato was not a mathematician but was evidently intrigued by mathematics. He often referred to it in his writings, promoted it in his school, the Academy, and greatly influenced not only the attitude of western philosophers toward it, but also the outlook of working mathematicians. His philosophy was centered on the belief that beyond the imperfect and unstable world of sense experiences there exists an ideal world of pure, perfect, and unchanging entities, which he called "ideas" or "forms". The particular objects that we experience in our material life are somehow modelled on or descended as imperfect copies from the ideal forms. The most important forms were goodness, truth, beauty, and other ideal entities of general ethical value, but Plato also often referred to mathematical objects as examples of forms, and hence to mathematical knowledge as an example of philosophical knowledge. In his political-philosophical dialogue, *Republic*, he advocated mathematics as an essential part of the education of an ideal leader, a "philosopher king" because, in addition to being useful in military operations, it turned "the soul upward, away from becoming and toward truth and being." According to Plato, mathematics led to truth and wisdom because it trained the mind to move from the experience of particular sensible objects to the irrefutable knowledge of abstract forms. For Plato, the objects of mathematics were not the material objects known through the senses, as for example this or that particular square field, or this group of five fingers or that group of five pebbles. They were abstract entities, as the square (the ideal geometrical figure that has four equal sides) and the number five. Only knowledge of such objects could be free of doubt and unchanging. Plato also believed that the cosmos is organized according to geometrical principles. In the *Timaeus*, he presented a story of creation according to which the physical elements, water, air, earth, and fire were constituted by different regular solids, which were in turn made by elementary triangles.[12]

Plato gave an example of mathematical knowledge in his philosophical dialogue, *Meno*, in which two characters, the philosopher Socrates and the philosophy student Meno debate the meaning of virtue. Socrates, Plato's teacher in real life and his literary persona, claims that anybody could find the truth and hence become virtuous.[13] To support his argument, he calls a slave boy and asks him to solve a geometrical problem: given a square of side 2 ft, what is the side of a square that is double the area? The boy's first answer is twice the original side, that is, 4 ft, but Socrates shows that this is incorrect because the area of a square of twice the side is four times the area of the original square. Then, Socrates guides the boy to the right answer. With the help of a geometrical diagram that stands for an ideal geometrical construction, he shows that the side of the square of double area is equal to the diagonal of the original square (fig. 4.11).

The discussion of the duplication of the square in *Meno* is the earliest instance of an extended and explicit mathematical argument. The fictional slave boy is not presented with a recipe-like series of instructions as were the pupils in the ancient scribal schools. Socrates guides him to use his reasoning to arrive at the solution. Another notable feature of the discussion is that while the question is posed numerically (the side of the original square is 2 ft) the answer is given only geometrically: the side of the double square

[12] See Chap. 6 for further discussion of Plato's ideas about mathematics and the physical world.

[13] Plato used the example to argue for his theory of recollection, according to which we do not discover new truths through reasoning but "recollect" knowledge that we possessed in our disembodied existence before birth.

is the diagonal of the original square. In other words, Plato moved from a measurement problem, very similar to the geometrical exercises in the Egyptian mathematical papyri and in the Babylonian tablets, to a kind of geometry in which measurement was no longer the goal. Although non-numerical, the answer was exact and irrefutable, and was reached purely by means of logical reasoning on an ideal construct. This was the kind of knowledge that for Plato exemplified science.

Sources

Aristophanes. *The Birds*, 414 BCE. Text available online at *The Internet Classics Archive*, http://classics.mit.edu/Aristophanes/birds.html

Asper, Mark. "The two cultures of mathematics in ancient Greece". In *The Oxford Handbook of the History of Mathematics*, edited by E. Robson and J. Stedall, 108–132. Oxford and New York: Oxford University Press, 2009.

Burnyeat, Myles Frederic. "Other Lives", *London Review of Books* 29, 4, 22 February 2007, 3–6.

Chemla, Karine. "Geometrical Figures and Generality in Ancient China and Beyond: Liu Hui, Zhao Shuang, Plato and Thabit ibn Qurra", *Science in Context* 18, 1, 2005, 123–166.

Cuomo, Serafina. *Ancient Mathematics.* London and New York: Routledge, 2001.

Euclid. *The Elements*, published online by D. E. Joyce at http://aleph0.clarku.edu/~djoyce/java/elements/elements.html

Høyrup, Jens. "Pythagorean 'Rule' and 'Theorem'—Mirror of the Relation between Babylonian and Greek Mathematics". In *Babylon: Focus mesopotamischer Geschichte, Wiege früher Gelehrsamkeit, Mythos in der Moderne*, edited by J. Renger, 393–407.2. Internationales Colloquium der Deustchen Orient-Gesellschaft. Saarbrücken: SDV Saarbrücker Druckerei, 1999.

Katz, Victor J. *A History of Mathematics. An Introduction.* 3rd Edition. Boston: Addison-Wesley, 2009.

Lloyd, Geoffrey Ernest Richard. *Magic Reason, and Experience. Studies in the Origin and Development of Greek Science.* Cambridge: Cambridge University Press, 1979.

Lloyd, Geoffrey Ernest Richard. *Demystifying Mentalities.* Cambridge, New York: Cambridge University Press, 1990.

Lloyd, Geoffrey Ernest Richard. "What was Mathematics in the Ancient World? Greek and Chinese Perspectives". In *The Oxford Handbook of the History of Mathematics*, edited by E. Robson and J. Stedall, 7–25. Oxford and New York: Oxford University Press, 2009.

Merton, Robert K. "The Matthew Effect in Science, II. Cumulative Advantage and the Symbolism of Intellectual Property", *Isis*79, 1988, 606–623.

Netz, Reviel. *The Shaping of Deduction in Greek Mathematics.* Cambridge: Cambridge University Press, 1999.

O'Connor, John Joseph, and Robertson, Edmund Frederick. "The Indian Sulbasutras". In *The MacTutor History of Mathematics Archive*, http://www-gap.dcs.st-and.ac.uk/history

Robson, Eleanor. *Mathematics in Ancient Iraq: A Social History.* Princeton and Oxford: Princeton University Press, 2008.

Russell, Bertrand. *History of Western Philosophy.* 2nd Edition. Florence, KY: Routledge, 2004.

Stigler, Stephen M. "Stigler's Law of Eponymy". Reprinted in *Statistics on the Table. The History of Statistical Concepts and Methods*, 277–290. Cambridge, MA and London, England: Harvard University Press, 1999.

Chapter 5

Hellenistic Mathematics, Euclid's *The Elements* in Particular

5.1 The Golden Age of Greek Mathematics

The classical age of ancient Greece (or Hellenic Age) came to an end in the second half of the fourth century BCE, when the Macedonian king Philip II defeated an alliance of city-states and took power over a unified pan-Hellenic state. Shortly thereafter, his son and successor, Alexander launched a series of military campaigns and created an empire that stretched from Egypt to the Indus River. Alexander's conquests spread Greek culture far and wide, and Greek became the language of educated people in the Mediterranean world. At the same time, the Alexandrian empire created the conditions for a mixing and cross-fertilization of different cultures. A new cosmopolitan world emerged, in which Greek culture was dominant but not exclusive. For this reason, this period is called the Hellenistic Age—Hellenistic means "Hellenized", or "Greekish". The major centre of Hellenistic culture was Alexandria of Egypt, a harbour city that Alexander founded in the Nile delta and that became the economic centre of the Mediterranean region. The socio-political environment of the Hellenic polis, which had fostered the emergence of demonstrative mathematics, no longer existed. Nonetheless, by this time mathematics was established as a special discipline of theoretical learning among the educated elite. It had acquired not only a body of knowledge that scholars deemed worth of systematizing, expanding, and preserving in writing, but also its own distinctive style of expression and communication. Furthermore, although education was still restricted to the privileged elite, science was encouraged by the new rulers because it served to give them a public image of splendour and refinement.

The most famous cultural institution of the Hellenistic Age was the Museum, a "temple of the Muses" that was founded in Alexandria of Egypt by Alexander's successor, the Egyptian king Ptolemy I Soter. Ptolemy I and his dynastic successors invited the best scholars of the time to live and work at the Museum under their royal patronage. Philosophers, poets, linguists, historians, geographers, engineers, and mathematicians gathered at the Museum, where they were lavishly supported so that they could pursue their studies and benefit from each other's lectures and conversations. They were of diverse origins and native tongues, and Greek was their common language. A large library was annexed to the Museum, the famous Library of Alexandria, which amassed the largest book collection in Antiquity. Although we do not know for sure, some sources report that at its peak the Library of Alexandria contained a half-million books. To give a modern analogy, the Museum was similar to an international research institute, with the difference that the mandate of the Museum was not to "expand the reach of knowledge for the benefit of humankind" or to "foster technological innovation for the advantage of society" (to paraphrase the mission statements of several modern day research institutions), but to increase the prestige and power of the king. A few of the Museum's scholars did engage in the design of engines and machines, but their inventions were intended to augment the military might of the state or to awe the masses with spectacular displays. For example, around 280 BCE the Museum scholars took part in a great parade organized and financed by Ptolemy II. The procession's theme was astronomical events. It was led by a poet, it included magnificent floats, among which was a giant automaton (a self-moving statue), and it was closed by infantry and cavalry forces. Not everybody was impressed by the feats of the royal scholars, however. At least one contemporary author reportedly described them as "well-fed bookworms, who quarrel without end in the Museum's birdcage." (Quoted in Cuomo 2001, 136.) The Museum remained a major centre of intellectual life for seven centuries, from its founding through the Roman conquest to the final decline of the Roman Empire. Mathematics, among other disciplines, flourished under royal patronage. The stars of Greek mathematics, Euclid, Archimedes, and Apollonius, lived in the Hellenistic Age and were variously associated with Alexandria. For this reason, this time is called the Golden Age of Greek mathematics. It is also from this time that the earliest mathematical works handed down to us in their entirety are dated. The first and foremost of them is *The Elements* by Euclid.

5.2 Who was Euclid, and What Exactly Did He Write?

That Euclid has come down in history as the author of the most famous math textbook is all we know for sure about him. A few educated guesses can be made from what later authors wrote about him. He probably studied at Plato's Academy in Athens and then moved to Alexandria of Egypt shortly after 300 BCE. He may or may not have led a school of geometry at the Museum. Archimedes, who lived in Sicily and was a little younger, probably studied in Alexandria and may have known Euclid. It is possible that Apollonius also studied in Euclid's school at the Museum. Euclid authored other books of geometry besides *The Elements*, as well as books in astronomy and optics, some of which have survived. Two stories have been handed down about Euclid. Both feature him as "the" teacher of geometry. In one, a student asks, "What will I get from learning these theorems?" As a reply, Euclid tells a slave to hand the student some change, "since he has to make a gain from what he learns."In the other, King Ptolemy asks Euclid to show him a shorter way to geometry than that laid out in *The Elements*. Euclid replies, "Sir, there is no royal road to geometry." Neither account can be confirmed, but we should not expect stories like these to give accurate biographical details. We should regard them as moral tales that convey how Euclid's followers perceived Euclidean geometry. The moral of the first is that Euclid's geometry is not something to be learned for material gain, and that a Euclidean teacher can rightly be dismissive of expectations of practical rewards. The second expresses a specific view of mathematical knowledge, where it comes from, how it is to be acquired, and who is best suited to possess it. It is a view very different from that expressed by the Old Babylonian depictions of the "rod and ring" ceremony, in which the king received the symbols of measurement (hence geometrical knowledge) from the hands of the god of justice. In that case, royalty was indeed portrayed as having a privileged access to geometry; in Euclid's case, on the contrary, no free pass is given for social status: even a king has to sweat his way through the textbook if he wants to learn.

While it seems beyond doubt that Euclid was the author of *The Elements*, there are open questions about the nature of his authorship. As most modern school manuals, the book was not an original work but a collection of mathematical knowledge accumulated by Euclid's predecessors. There is indirect evidence that Greek theoretical mathematicians had begun producing textbooks in the fifth century BCE, when theoretical mathematics became a teaching subject in the liberal education of the upper classes. It is reported that the author of the first of these teaching manuals was Hippocrates of Chios, who worked and taught in Athens, and is also famous for proving that two circles are to one another as the squares of their diameters and for finding ways to calculate the areas of curved figures.[1] *The Elements* was a common title, and it is possible that Euclid used previous works of this kind, or parts of them, to compose his own book. At any rate, his book was so successful that it made the others appear no longer necessary. Scholars of the following generations trusted Euclid's text to be sufficiently comprehensive. Therefore, they read, taught, copied, and transmitted it, while letting the works from which Euclid had taken his content fall into neglect and be lost forever. That is perhaps the main single reason why so little direct evidence of pre-Euclidean theoretical mathematics has arrived to us. Furthermore, Euclid's text is very opaque regarding its sources. The concept of authorship and the publishing conventions of the time did not require an author to cite his sources or make clear what parts of a text were reported and which were original. There existed no copyright or other legislation that protected intellectual property. In addition to this, the expository style chosen by Euclid was very austere and self-restrained. It left no room for personal or historical commentary, remarks on the composition of the book, or sidelights. As a result, while it is evident that most of the content of *The Elements* is not an original creation, we have to rely on fragmentary and indirect evidence from other authors to reconstruct the prequel to *The Elements* and the contributions that Euclid made when he compiled what he had learned from his predecessors.

Recovering the original text written by Euclid is also a problem. The earliest copy of the Greek text that has survived dates from 888 CE, that is, over a millennium after Euclid wrote it.[2] This and the other pre-modern copies that have survived, however,

[1] See Sect. 4.6.

[2] This copy, known as MS D'Orville 301, was made in Constantinople (modern-day Istanbul) and is now kept in the Bodleian Library in Oxford, England. See http://www.claymath.org/library/historical/euclid/. Modern English translations can be found online in the public domain, for example an edition by David Joyce that includes Java applets, at http://aleph0.clarku.edu/~djoyce/java/elements/elements.html, and one by Richard Fitzpatrick, which includes Heiberg's Greek edition, at http://farside.ph.utexas.edu/euclid.html.

all differ from one another in significant ways, so that none of them can be trusted to reproduce the original with perfect accuracy. The Greek text that is currently used for modern translations is the result of a painstaking reconstruction by a nineteenth-century philologist, Johannes L. Heiberg, who collated different Greek and Latin versions of the book. The reproduction of a book was a difficult task before the invention of printing. Not only did errors creep into the text with successive manual copying, but also comments added by readers on the margins of the pages called scholia, were often mistakenly incorporated into the text. Moreover, editors and commentators did not have a modern conception of authorship, which demands respect for the integrity of an intellectual work. While for literary works they did take some care to preserve the exact language and form of the original, for mathematical works they deemed it permissible to insert what they thought were necessary additions, clarifications, and simplifications. The textual tradition that carried *The Elements* down to us across 23 centuries is intricate, and we may never be able to completely restore Euclid's original text (fig. 5.1).

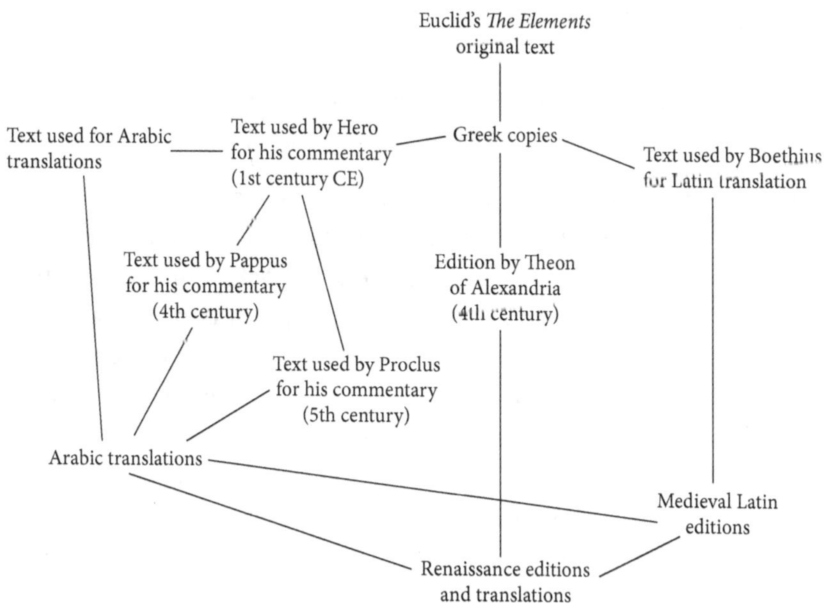

Fig. 5.1. The pre-modern textual tradition of Euclid's *The Elements*.
Adapted from Cuomo, *Ancient Mathematics* (2001), 127.

5.3 Euclid's *Elements*

The text of Euclid's *Elements* that has arrived to us deals with plane and solid geometry, the theory of proportions, and the theory of numbers. It is composed of thirteen parts, which are traditionally called "books" but are rather more like chapters than books in the modern sense. The exact subdivision is as follows:

- Books I to VI: plane geometry
  - I: First principles (definitions, postulates, and common notions), triangles, parallel lines, Pythagoras theorem
  - II: transformation of areas
  - III: circles
  - IV: regular polygons
- Books V and VI: theory of proportions
- Books VII to IX: theory of numbers
- Book X: theory of irrational magnitudes
- Books XI to XIII: solid geometry
  - XI: parallelepipeds
  - XII: areas and volumes of solids, method of exhaustion
  - XIII: the five regular polyhedral, or "Platonic solids"

According to tradition, the theorems of the first two books and the theory of numbers are based on the work of the Pythagoreans. Parts of Books III and IV draw on the work of Hippocrates of Chios. The theory of proportions in Books V and VI, and the method of exhaustion, by which areas and volumes are calculated in Book XII, are attributed to Eudoxus of Cnidus, a mathematician and astronomer (there was no such distinction in Antiquity) of Plato's school. And the theory of irrational magnitudes in Book X and the construction of the five regular polyhedra, or "Platonic solids", are the work of Theaetetus, also from Plato's Academy.[3] Besides collecting and ordering this material, Euclid contributed some of the content. In particular, it is believed that the theory of parallel lines and the proof of the so-called Pythagorean theorem that he presents in Book I were his own.

[3] See Chap. 4.

Most importantly, however, Euclid ordered the material and organized it according to a specific logical structure. Following Aristotle's views on scientific discourse, he transformed the body of Greek theoretical mathematics into an *axiomatic deductive system*, that is, a system of interrelated propositions that derive with logical necessity from a set of first principles. The first principles are indifferently called axioms or postulates in modern terms, but Aristotle distinguished between axioms, or common notions, and postulates, or special notions. The former are general to all sciences, while the latter belong to a specific science. Accordingly, the first principles that Euclid placed at the start of his logical system consisted of five postulates and five common notions. The Heiberg edition also includes twenty-three definitions at the beginning of Book I, and more definitions at the beginning of the remaining books. The rest of *The Elements* is composed of 465 propositions, that is, theorems, such as "The sum of the interior angles of a triangle equals two right angles", and problems, such as "How to draw a straight line through a given point parallel to a given line", followed by their proofs. Each proof ends with a closing formula, "as it was to be demonstrated" (in Latin editions the abbreviation *QED* was used, for *Quod Erat Demonstrandum*) for a theorem, and "as it was to be done" for a problem (in Latin, *QEF*, for *Quod Erat Faciendum*).

The propositions are arranged in strict deductive order: each proposition is proven solely on the basis of the preceding propositions and of the first principles. For example, the proof of Proposition I.1 (Proposition 1 of Book I) requires some of the first principles only, that of Proposition I.2 some of the first principles and Proposition I.1, that of Proposition I.3 some of first principles and Proposition I.2, and so on. The second-last proposition of Book I, Proposition I.47, which is the so-called Pythagorean theorem, requires some of the first principles and Propositions I.4, I.14, I.31, I.41, and I.46; Proposition I.46 requires Propositions I.3, I.11, I.29, I.31, and I.34, and so on. Although, as we will see, modern mathematicians have found flaws in the arrangement devised by Euclid, his system has been the most admired and imitated model of a logically organized demonstrative discourse for over two millennia. A modern author opined that, "to the modern reader [Euclid's] work is incredibly dull. There are no examples; there is no motivation; there are no witty remarks; there is no calculation. There are simply definitions, axioms, theorems, and proofs." (Katz 2009, 51.) Yet, *The Elements* has been an inspiration to countless readers across the ages, from Thomas Hobbes to Abraham Lincoln to Albert Einstein, and has been held as a model to imitate not only in mathematics and science but also in any field where disciplined reasoning is valued. Many have also found aesthetic pleasure in the austere elegance of Euclid's construction.

Euclid's Definitions

The following is a sample of Euclid's initial definitions:

1. *A point is that which has no parts*
2. *A line is a breadthless length*
3. *The extremities of a line are points*
4. *A straight line is a line that lies evenly with the points on itself*
5. *A surface is that which has length and breadth only*

...

23. *Parallel straight lines are lines which, being in the same plane and being produced indefinitely in both directions, do not meet one another in any direction*

Other definitions appear at the beginning of individual books. For example, at the start of Book VII, the first of the books on the theory of numbers, we find 22 definitions concerning numbers:

1. *A unit is that by virtue of which each of things that exist is called one*
2. *A number is a multitude composed of units*
3. *A number is a part of a number, the less of the greater, when it measures the greater*

...

11. *A prime number is that which is measured by a unit alone*

...

With the professionalization of mathematics in the nineteenth century, mathematicians became concerned with grounding their discipline on undisputable logical foundations. They subjected Euclid's work to a thorough examination, and found that below its admirable structure there was room for improvement. Much of their criticism targeted the definitions, which they found to be at best vague and at worst useless. The problem is that they are a hodgepodge of claims, some of which attempt to identify the objects of mathematical discourse while others describe their properties. They are meant to be starting points, but many of them rely on notions that are not simpler than the defined terms, as for example "breadthless length" and "to lie evenly", and that remain undefined. In some cases, they are tautological, as in the case of Definition VII.1, "A unit is that by virtue of which each of the things that exist is called one." In the case of geometrical objects, they point to a mental picture of geometrical objects that can only derive from physical experience. More specifically, they describe idealizations and abstractions of certain kinds of material objects. It has been argued, for example, that the idea of "breadtheless length" is an abstraction from an older understanding of lengths as having a given thickness, as is seen for instance in the geometrical problems in Old Babylonian tablets. Similarly, the idea of a line "lying evenly" with some points on it may represent the ideal result of the practice of drawing straight lines with the help of sighting instruments in construction and land surveying. Since the nineteenth century, mathematicians have moved away from a conception of geometry as a theory of physical space and have rejected the need to ground the first principles in sense experience. They have accepted that an axiomatic deductive system must include a set of primitive terms, which cannot be defined but acquire meaning through their being used according to the properties defined by axioms and postulates.

However imperfect, Euclid's definitions do serve the purpose of conveying to all readers the same mental pictures of points, lines, surfaces, and other mathematical objects. If they appear inadequate, it is probably because modern standards of logical rigour are being applied retroactively to something that Euclid never actually intended to be a complete inventory of rigorous definitions. In the original text they did not appear as a numbered list, as they do in Heiberg's edition, but as a continuous piece of prose. They constituted pithy introductions to the chapters, shorter than the more expansive introductions that are found in books by Archimedes, Apollonius, and other Greek mathematicians but similar to them in their purpose. The introductions of mathematical books were, as one historian put it, "the place where mathematicians meet with non mathematicians". (Netz 1999, 96.) Their function was to introduce the reader to the objects of mathematics in a non-technical way, not to fix completely and unambiguously the technical meaning of the terms. In fact, as modern mathematicians realized, mathematical discourse itself could proceed without definitions.

Euclid's Axioms and Postulates

Euclid's five common notions, also called axioms, are,

1. Things that are equal to the same thing are also equal to one another
2. If equals are added to equals, the wholes are equal

3. If equals are subtracted from equals, the remainders are equal
4. Things that coincide with one another are equal
5. The whole is greater than its parts

As we have seen, Aristotle had already called for the need to recognize explicitely such basic assumptions, which he considered self-evident ideas, as starting points of demonstrations, and some of his examples are identical to Euclid's. Over time, the distinction between "axiom" and "postulate" has disappeared, and today these two terms are used as synonyms, but the need to spell out the premises of a logical argument has only been confirmed.

The five geometrical statements chosen by Euclid as postulates for his system are,

1. A straight line can be drawn from any point to any other point
2. A finite straight line can be produced continuously in a line
3. A circle may be described with any center and distance
4. All right angles are equal to one another
5. If a straight line falling on two straight lines makes the interior angles on the same side less than two right angles, then the two straight lines, if produced indefinitely, meet on that side on which are the angles less than two right angles

These first principles also have been the object of criticism. According to modern axiomatic theory, the axioms must be consistent, that is, they must not contradict one another. Furthermore, they must be complete, that is, they have to express *explicitly everything* that is needed to demonstrate the propositions. In other words, the theory must not contain any tacit assumption. Finally, the axioms must be independent, that is, it must be impossible to derive any one of them from the others. After over two thousand years of studies, no one has found a contradiction in Euclid's first principles; however, mathematicians have found many hidden assumptions in Euclidean geometry, and for a long time it was suspected that the fifth postulate was not independent from the other four.

Concerning the issue of tacit assumptions, let us consider, for example, Proposition I.1: "To construct an equilateral triangle on a given finite straight line." Euclid constructs an equilateral triangle by joining the extremities of the given line with the point of intersection of two circles, having each one of the extremities as the centre and the line as the radius (see fig.5.2).

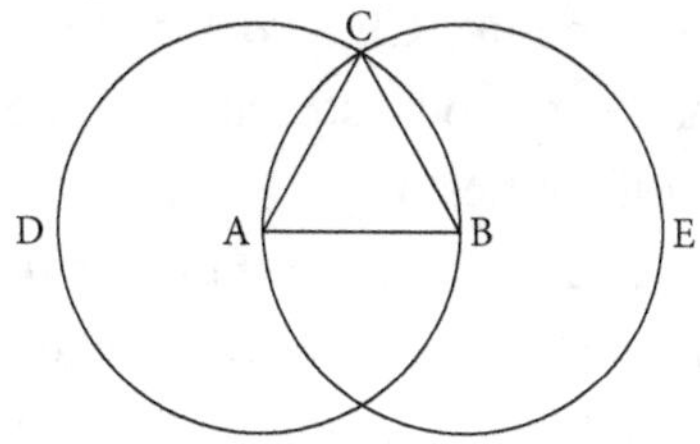

Fig. 5.2. Diagram of proposition I.1 of Euclid's *Elements*.

The demonstration uses Definitions I.15 and I.20 (circle and equilateral triangle), Postulates 1 and 3, and Common Notion 1, but it also takes for granted that the two circles do have a point of intersection. In other words, it makes a tacit assumption about the continuity of the circumferences. This assumption may appear obvious from our intuitions about circles, and it seems supported by the diagram. Nonetheless, it cannot be deduced from the first principles, and therefore it must be considered an additional starting point of the demonstration which is not explicitly mentioned as a first principle. Euclid's argument for equality of geometric figures also relies on the possibility to *superpose* two figures and make them coincide. Euclid tacitly assumes that it is possible to transport one figure through space and bring it over another without any change in the displaced figure. But there is nothing in the first principles that guarantees this to be possible. For this reason, the British philosopher Bertrand Russell called Euclid's proof of Proposition I.4, "If two triangles have two sides respectively equal and the angles between them equals, then [...] triangle equals triangle" (the SAS criterion of congruence in modern geometrical parlance), "a tissue of nonsense." He concluded that the value of *The Elements* "as a masterpiece of logic has been very grossly exaggerated." (Russell 1902, 167.) Such judgmental retrospections are open to the charge of being historically insensitive, and in fact Euclid's defenders responded sarcastically that Euclid's only fault had been that of not having studied Russell. All the same, many other hidden assumptions were uncovered by geometry experts. In 1899, the German mathematician David Hilbert published *Foundations of Geometry* (*Grundlagen der Geometrie* in German), a revision of Euclidean geometry that included the first complete set of first principles. It consisted of twenty axioms and three primitive terms.

A different strain of criticism concerned the independence of the postulates, and targeted Euclid's fifth postulate. Already some of the first students of *The Elements* noticed that this postulate did not appear to be as self-evident as the other four, and that it was not used in the first 28 propositions. They argued that it was not independent of the first four postulates, that it should be possible to derive it from them. Therefore, it should have been regarded as one of the propositions. In the fifth century CE, Proclus wrote in his commentary to *The Elements* that the fifth postulate "ought to be struck from the postulates altogether, for it is a theorem involving many difficulties, which Ptolemy in a certain book set himself to solve, and requiring for its demonstration a number of definitions as well as theorems." According to Proclus, then, attempts to prove Euclid's fifth postulate on the basis of the other four had been underway since at least the time of Claudius Ptolemy, that is, the second century CE.[4]

Let us take a closer look at what was involved in the issue. Euclid's fifth postulate became known as the "parallel postulate" because it supported Euclid's theory of parallel lines. Euclid defined parallel lines as follows:

> Definition I.23: Parallel lines are straight lines which, being in the same plane and being produced indefinitely in both directions, do not meet one another in any direction.

It should be noted first of all that Euclid's definition does not guarantee that parallel lines exist. The possibility of lines "being produced indefinitely", in fact, was assured by the third postulate:

> Postulate 3: A straight line can be produced continuously in a line.

Yet, this postulate did not state that the lines can be extended to infinity in an open way, that is, in such a way that they never return onto themselves. The possibility of *infinite* extension was one of the hidden assumptions of Euclid's theory. Only in the second half of the nineteenth century, in the wake of the discovery of the first non-Euclidean geometry, did mathematicians notice that this assumption was equivalent to a tacit assumption about the nature of the space in which geometrical figures such as straight lines were taken to exist.[5] In the 1850s, the German mathematician Bernhard Riemann realized that it was possible to formulate a second kind of non-Euclidean geometry, in which space was "curved", the assumption of infinite extension was denied, and no parallel lines existed (Riemannian geometry).

Until the nineteenth century, everybody assumed that parallel lines existed, and the fifth postulate stated a condition under which two lines were not parallel (fig. 5.3):

[4] See Chap. 6 for a discussion of Claudius Ptolemy.

[5] The discovery of the first non-Euclidean geometry is briefly discussed below.

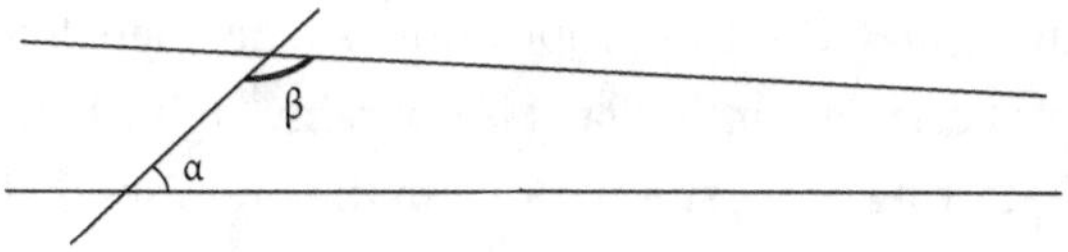

Fig. 5.3. Euclid's fifth postulate, a.k.a. "the parallel postulate".

> Postulate 5: If a straight line falling on two straight lines makes the interior angles on the same side less than two right angles, then the two straight lines, if produced indefinitely, meet on that side on which are the angles less than two right angles.

One observation was that the fifth postulate was not used to prove the first 28 propositions. A considerable portion of geometry simply did not need it. Moreover, it was not as pithy as the other four, nor did it appear to be as self-evident. Another observation was that it was closely related to the 27th and 28th theorems of the first book of *The Elements* (fig. 5.4):

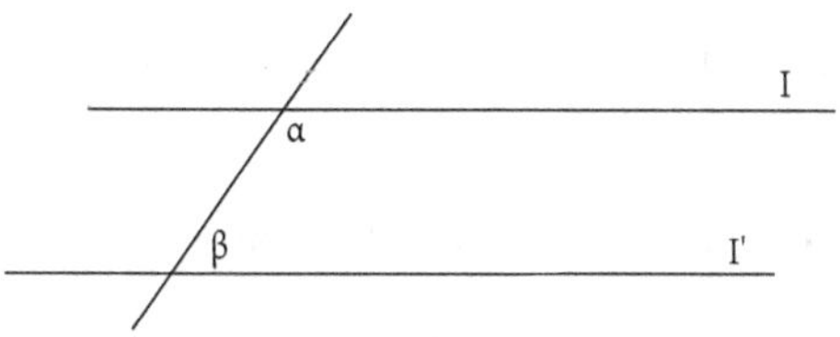

Fig. 5.4. Proposition I.27 and I.28 of *The Elements* state conditions under which two lines are parallel. They do not require the fifth postulate to be proven.

> Proposition I.27: If a straight line falling on two straight lines makes the alternate angles equal, then the two lines are parallel to one another.

> Proposition I.28: If a straight line falling on two straight lines makes the exterior angle equal to the interior and opposite angle on the same side, or the sum of the two interior angles on the same side equal to two right angles, then the two lines are parallel to one another.

These two propositions state conditions under which two lines are parallel, and they are proven solely on the basis of the first four postulates, without involving the fifth. Finally, the fifth postulate is the inverse of the second part of Proposition I.28 (namely, if the sum of the interior angles is equal to two right angles, the lines are parallel), for it derives the opposite conclusion (the lines are *not* parallel) from the opposite hypothesis (if the sum of the interior angles on the same side is *not* equal to two right angles). Several readers of Euclid were then led to argue that the proposition that Euclid used as the fifth postulate ought to be derivable as a theorem on the basis of the first four postulates. They did not succeed in coming up with a proof, however. The reason resides in logic: a proposition and its inverse are logically independent from one another. A simple example to consider is the proposition, "If it is rainy, I carry an umbrella". If this proposition is true, there follows logically the contrapositive, "If I do not carry an umbrella, it is not rainy." It is a logical fallacy, however, to conclude from the starting proposition (or from the premises that make the starting proposition true) that also the inverse, "if it is not rainy, I do not carry an umbrella" follows. The inverse is logically independent from the starting proposition. It has either to be accepted without proof or proven from additional premises. Unlike his critics, Euclid understood that the condition of non-parallelism could not be derived from the first four postulates but had to be stated as an independent additional postulate.

From the fifth postulate, Euclid derived a series of theorems. He proved two more theorems about parallel lines:

> Proposition I.29: A straight line falling on parallel lines makes the alternate angles equal to one another, the exterior angle equal to the interior and opposite angle, and the sum of the interior angles on the same side equal to two right angles.
>
> Proposition I.30: Two lines parallel to the same line are parallel to one another

From these, there followed that given a line and point not on the line, one and only one parallel to the line could be drawn. Being logically equivalent to Euclid's fifth postulate, this proposition was proposed as a more intuitive formulation of the postulate by the Scottish mathematician John Playfair in the eighteenth century, and for this reason it is today known as the Playfair Axiom.

> Playfair Axiom: given a line and point not on the line, there exist one and only one parallel to the line.

Then, Euclid proved the well-known theorem about triangles (fig. 5.5),

> I.32: In any triangle, if one of the sides is produced, then the exterior angle equals the sum of the interior and opposite angles, and the sum of the interior angles is equal to two right angles

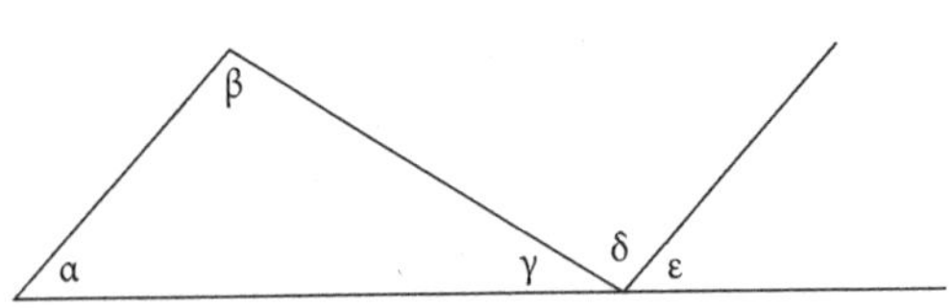

Fig. 5.5. Proposition I.32 of *The Elements*. It requires Euclid's fifth postulate or an equivalent postulate, such as the Playfair Axiom, to be proven.

Thanks to his theory of parallel lines, Euclid could prove a series of theorems on equal triangles and parallelograms, and finally arrived at the so-called Pythagorean theorem:

> Proposition I.47: In right-angled triangles the square on the side opposite the right angle equals the sum of the squares on the sides containing the right angle.

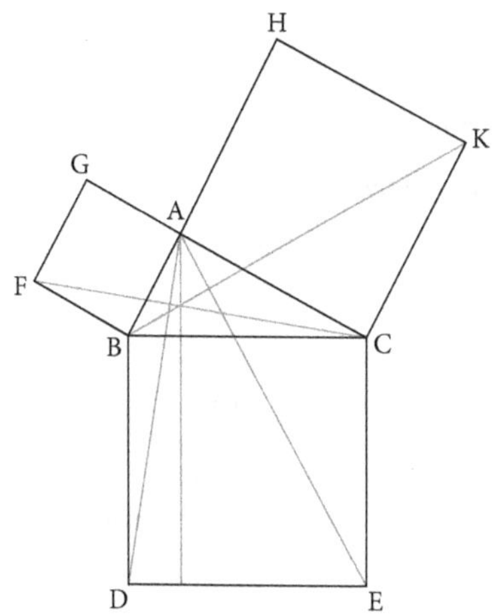

Fig. 5.6. Proposition I.47 of *The Elements,* a.k.a. the Pythagorean theorem. It depends on the fifth postulate for its proof.

Many students of *The Elements* across the ages were awed by the logical edifice erected by Euclid. There were also a few who found it unnecessarily convoluted and impenetrable. For example, the nineteenth century German philosopher Arthur Schopenhauer, who believed that geometry should be more intuitive than Euclid's ironclad but opaque logic, complained (Schopenhauer, *The World as Will and Representation*, 1818):

> Often, as in the proposition of Pythagoras, lines are drawn, we don't know why, and it afterwards appears that they were traps which close unexpectedly and take prisoner the assent of the astonished learner.

Well known properties of parallel lines, triangles, and parallelograms, which every mathematician believed to be indubitable, depended on the fifth postulate for their proofs. Over more than twenty centuries, several mathematicians questioned Euclid's logical construction, and tried their hand at improving it by demonstrating that the fifth postulate was truly a theorem provable on the basis of the first four postulates. They did not succeed, yet their efforts were not unproductive. Eventually, in the 1820s, the German

Carl Friedrich Gauss, the Hungarian Janos Bolyai, and the Russian Nikolai Lobachevsky came to a double realization. Euclid was right but, at the same time, his geometry is not the only possible geometry. Alternative geometries can be formulated without the fifth postulate, and they are as logically valid as Euclid's. They are called non-Euclidean geometries.

Sources

Calinger, Ronald. *A Contextual History of Mathematics.* Upper Saddle River, NJ: Prentice Hall, 1999.

Cuomo, Serafina. *Ancient Mathematics.* London and New York: Routledge, 2001.

Katz, Victor J. *A History of Mathematics. An Introduction.* 3rd Edition. Boston: Addison-Wesley, 2009.

Netz, Reviel. *The Shaping of Deduction in Greek Mathematics.* Cambridge: Cambridge University Press, 1999.

Russel, Bertrand. "The Teaching of Euclid", T*he Mathematical Gazette* 2, 33 (1902), 165–167.

Saito, Ken. "Reading ancient Greek mathematics". In *The Oxford Handbook of the History of Mathematics*, edited by E. Robson and J. Stedall, 801–826. Oxford and New York: Oxford University Press, 2009.

Chapter 6

Mathematics and the Physical World in Antiquity

6.1 Mathematics and the World of Experience in Ancient Greece

The emergence of theoretical mathematics in ancient Greece opened up the question of the relationship between mathematical knowledge and the physical world. As long as mathematics was simply a cluster of practices related to measuring and counting, its relation to the world of experience was straightforward: mathematical reasoning and procedures were useful tools in everyday problem-solving. When Greek thinkers recognized that numbers and geometrical figures were not physical objects but abstract entities, however, they started wondering about their "reality" and debating the purpose of studying them. Everyone agreed that axiomatic-deductive mathematics—provided that one was able to identify the right set of axioms—was the most persuasive kind of knowledge and the paradigm of rational thinking. Opinions differed, however, on how such abstract knowledge related to the phenomena that we experience through the senses. Broadly speaking, the views on this issue fell into two camps, the Platonist and the Aristotelian. The Platonist view was based on Plato's theory of "forms". Plato and his followers believed in the existence of two levels of reality, one empirical and the other ideal.

We experience one level of reality through the senses, but the knowledge we acquire in this way can only be incomplete, inexact, and uncertain. On the other level, there exists a world of entities—"forms" or "ideas" in Platonic language—that are not sensible but are fully knowable through reason. For a Platonist, mathematical objects as, for example, *the* square and *the* number three were forms and existed on the ideal level as a "things in themselves", perfect, unchanging, eternal, and independent of the particular physical objects that more or less resembled them, as for example a square tile or three fingers. Plato recognized the usefulness of practical mathematics; he noted, for instance, that arithmetic and geometry would be of use to a military leader. Nonetheless, he argued that the leaders of the ideal republic should study mathematics not "like merchants and retailers, with a view to buying and selling," but as philosophers who aim at grasping the ideas of beauty, justice, and the good. They should be educated in the mathematical sciences for the purpose of elevating the soul, turning it "from the world of becoming to that of truth and reality." In the same discussion, Plato argued also that the ideal rulers should learn astronomy, not because "the observation of the seasons and of months and years is as essential to the general as it is to farmers and sailors", but because astronomy was a branch of mathematics and thus shared the mind-elevating power of this science. Hence, a true astronomer should not become mired in the observation of "the starry heaven which we behold" but rather aim at the knowledge of "the true realities of velocities", which are found "in pure numbers and perfect figures" (Plato, *The Republic*, ca. 375 BCE.) In some of his works, Plato suggested that sensible objects, which are flawed and fickle, are imperfect copies or shadows of the forms, which are unchangeable, timeless, and invisible to the eye but intelligible to the mind. He and his school adopted the doctrine of the Pythagoreans, according to which underneath all the apparent irregularity and disorder the physical world is structured according to mathematical principles of proportion and symmetry. In this view, a philosopher ought to be concerned with phenomena, but only to get beyond their imperfections and deceptions to recover the forms after which they are modelled. Plato gave the clearest presentation of this view in his book *Timaeus*, a creation story according to which the physical world was created by a "Divine Craftsman" (or "Demiurge") as a harmonic whole governed by geometry and mathematical proportions. For example, the Divine Craftsman made the four physical elements, earth, water, air, and fire, out of particles shaped like regular polyhedra, which he made out of triangles. Plato's story was an allegory that left open the question of how immaterial forms could have generated physical objects. Despite this difficulty, the

Platonic view that mathematical principles could be used to explain the features of the physical world has been highly influential from Antiquity up to the Modern Age.

In contrast, Aristotle and his followers believed that the world of experience constituted the one and only reality. For them, mathematics was not the study of a special class of non-physical entities. It was concerned with all objects, physical or otherwise, but only in regard to their mathematical attributes. Arithmetic dealt with any kind of countable objects, but limitedly to their being countable, and geometry dealt with any kind of spatial objects, but limitedly to their shapes, sizes, and spatial relations. Both Plato and Aristotle regarded mathematical entities, such as numbers and shapes, as abstract entities, but Aristotle's understanding of abstractness was different from Plato's. For Aristotle, abstractions were not things in themselves but derived from a mental process of stripping away physical qualities, such as hotness or coldness, hardness or softness, wetness or dryness, and so on, from physical objects, until only the quantity, extension, and shape of the body remained. The Aristotelians regarded mathematics as a science that was very general, in the sense that it applied to everything, yet very limited in its ability to explain the nature of things. More specifically, Aristotle defined nature ("*physis*") as the "principle of change", which he saw as residing in the essential qualities of substances. The study of nature and change was *physics*, a theoretical science concerned with finding the causes of phenomena through an examination of the essential qualities of the substances involved. Hence, physics was clearly distinct from mathematics. The Aristotelians admitted, however, that physics and mathematics overlapped in a few special fields of study, namely, astronomy, harmonics, optics, and mechanics. In astronomy, for example, the nature of celestial bodies was revealed through the shapes and sizes of their motions.

6.2 Astronomy, Celestial Phenomena, and "Plato's Problem"

Astronomy has been practised since time immemorial by many cultures around the world. To give just a few examples, the earliest known mathematical artifacts are lunar calendars; Neolithic monuments like Stonehenge in southern England appear to have been aligned according to the positions of the Sun; and clay tablets from the first millennium BCE show that in Babylonia there existed a class of civil servants specialized in astronomy. For centuries, despite the rise and fall of several empires and dynasties, these specialists systematically made and recorded observations of positions of the stars, the Sun, the Moon, and the five planets visible with the naked eye, Mercury, Venus, Mars, Jupiter, and Saturn. Starting in the sixth century BCE, the Babylonian astronomers were able to use their institutional body of data to develop arithmetical algorithms to calculate the future occurrence of specific astronomical events. As far as we know, however, the Greeks were the first to formulate geometrical models of astronomical phenomena. The word "phenomenon" (plural, "phenomena") derives from the ancient Greek verb *phainesthai* ("to appear") and means "that which appears" or "that which is observed". While in modern usage it is sometimes used to describe an exceptional or inexplicable happening (as in "paranormal phenomenon",) in Antiquity it referred to ordinary, repetitive natural occurrences that were deemed amenable to rational explanation. The foremost phenomenon of all, the first that Greek astronomers sought to explain, was the diurnal rotation of all the bodies in the sky ("diurnal" comes from *dies*, the Latin word for "day".) The sun rises daily in the east, describes an arc in the sky, and sets in the west, determining the recurrence of day and night. At night-time, the stars are seen to describe similar circular paths, either rising above the horizon and setting below it in the southern sky (for us in the northern hemisphere,) or completing circles around a fixed point in the northern sky called the celestial north pole, while maintaining their reciprocal positions unchanged. The duration of one full rotation constitutes the day, the basic unit of time in all known cultures. The next evident phenomenon is the annual cycle, the cycle of the seasons ("annual" comes from *annus*, the Latin word for "year".) The Sun rises exactly in the east and sets exactly in the west only once in the spring and once in the fall. These are the only days in which the duration of daylight is equal to that of the night time, and for this reason they are called equinoxes (from the Latin *aequa* and *nox*, "equal" and "night"). After the spring equinox, the points of sunrise and sunset shift toward the north and the daily arc of the Sun above the horizon becomes higher, until it reaches a maximum height. This is the time of the shortest night, and is called summer solstice. Afterward, the Sun reverses its shifting, reaches the fall equinox, and continues to a point at which the day is shortest and the night longest—the winter solstice. From there, daytime starts increasing again until it returns to the spring equinox. The duration of one full season cycle constitutes a year, another common time unit (fig. 6.1).

The lunar cycle, that is, the recursion of the phases of the Moon (new moon, waxing moon, full moon, and waning moon) is another major astronomical phenomenon, and it also provides a common time unit, the month. While each of these astronomical cycles

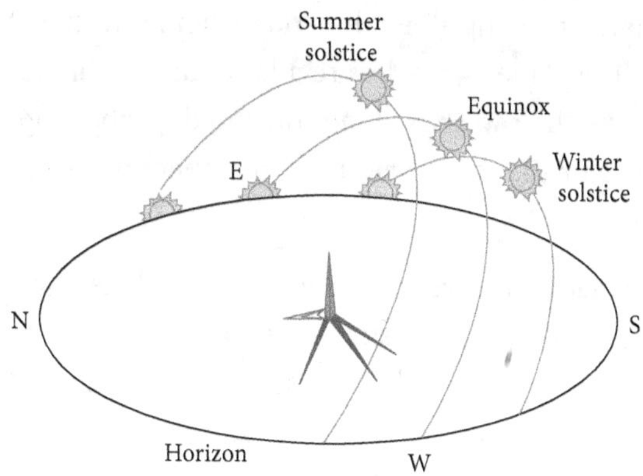

Fig. 6.1. The annual cycle of the Sun, as observed by means of the shadows of a sundial.

presents an evident regularity, their relations are not straightforward. The lunar month does not always contain exactly the same number of days, and the solar year does not contain an exact number of days or an exact number of lunar months. Finding ways to combine day, month, and year in an orderly calendar for agricultural, religious, and political purposes was perhaps the main motivation behind the early practice of astronomy.

Early astronomers noticed that the stars keep their reciprocal positions in the daily cosmic carousel, and form stable constellations; for this reason, they called them "fixed stars". Seeking to determine the occurrence of equinoxes and solstices, however, they observed that the Sun changes its position relative to the fixed stars. Therefore, many stars are visible only for part of the year, when the Sun is sufficiently distant from them not to obscure them with its light. The Egyptians, for example, knew that the appearance of the Isis star, known today as Sirius, just before dawn announced the annual flood of the Nile. The Babylonian astronomers invented methods to measure the position of the Sun relative to the fixed stars. They recorded that in every annual cycle the Sun passes by the same set of twelve constellations, which they named the Zodiac. Although they did not formulate a geometrical model, they described the motion of the Sun as a circular "travel" along the Zodiac in the direction opposite to that of the diurnal motion. They invented the sexagesimal system of units for measuring angles that we still use today, and recorded the positions of the Sun in degrees within the sectors corresponding to the constellations of the Zodiac, with each sector spanning 30° (in astronomical observations, the positions of celestial bodies are measured in terms of angles between the line of sight and a direction of reference.)

The Babylonians were also interested in five "wandering stars", which are known to us today as the planets (from the Greek *planetes*, "wanderers"). Like the Sun, the planets were observed to move among the fixed stars, traveling along the Zodiac not far from the path of the Sun but each at its own speed. The planets were less regular than the Sun, for at certain times they were observed to reverse their westward motion and describe loops on the backdrop of the fixed stars before returning to their track. This phenomenon is called the retrograde motion of the planets.

The interest of ancient astronomers in calculating the exact positions of the planets was related to astrology, the belief that the configurations of celestial bodies at specific times, for instance at the time of someone's birth, determined the course of human life and human affairs. Astrology originated in Babylonia around 500 BCE and spread to the Greco-Roman world around 300 BCE. In modern times it has been shown to be devoid of scientific foundations and has been excluded from the science of astronomy. Today it survives only as folklore or as a fringe belief. In Antiquity, however, astrology and astronomy were two aspects of the same discipline, and the practices that we today identify with astronomy were closely connected with the practice of astrological forecast. Other phenomena that captured attention were, of course, the solar and lunar eclipses, which were believed to be omens of important events on Earth. To achieve the ability to predict them was another motivation for ancient astronomers.

From the fourth century BCE on, Greek philosophers endeavoured to account for the astronomical phenomena by assuming that we stand on a spherical Earth placed at the centre of a finite, spherical cosmos. They supposed that all the celestial bodies were affixed to the inner surface of an invisible giant sphere, which revolved from east to west around an axis passing through the Earth's poles, completing one revolution a day. This two-sphere model was sketched philosophically by Plato and developed mathematically by Eudoxus of Cnidos, who studied and worked for some time at the Academy. Unfortunately, Eudoxus's works are lost and we know them only from second-hand reports. It is clear, in any case, that the two-sphere model constituted the basic theoretical framework of Greek mathematical astronomy and enabled astronomers to form models of other astronomical phenomena as well. Mapping the changing position of the Sun onto the sphere of the fixed stars made it possible to explain the annual cycle as produced by a west to east revolution of the Sun along a circle, called the ecliptic, inclined at an angle with respect to the equator (fig. 6.2). The obliquity of the ecliptic, that is, the angle it formed with the equator, was determined by measuring the difference between the altitudes of the Sun at noon at the summer and winter solstices. Around 300 BCE, Euclid knew it to be equal to "one-fifteenth of the circle". In the following century, the Greek astronomers adopted the angle units of the Babylonians and expressed this value as 360°/15 = 24°. In the second century CE, Claudius Ptolemy refined the measurement to 23° 51' 20". This value differs from the modern value by about 5°. The difference is due in part to a measurement error of Ptolemy and in part to the fact that the inclination of the ecliptic has decreased by about ¼°.

The two-sphere model allowed the observed motion of the Sun to be pictured as composed of two motions, the daily rotation of the celestial sphere and the yearly revolution along the ecliptic. Eudoxus accommodated this in his spherical model by adding a third sphere, still centered on the Earth, but to which only the Sun was attached. The sphere of the Sun rotated together with the sphere of the stars while at the same time it rotated independently and more slowly in the opposite direction around an axis inclined with respect to the north-south axis (fig. 6.3).

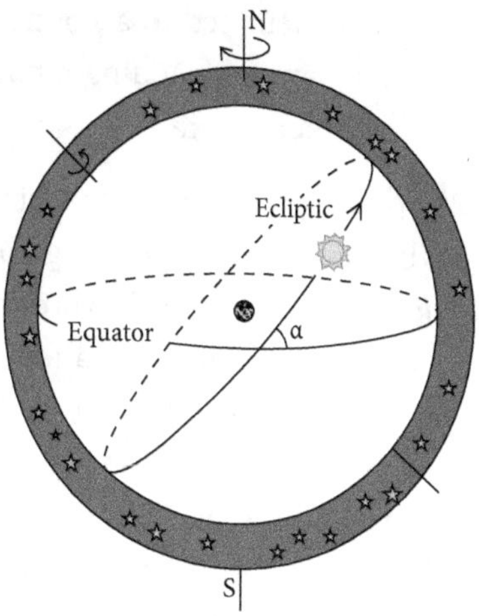

Fig. 6.2. Eudoxus's concentric spheres model, including the sphere of the stars and the sphere of the Sun. In this diagram, the obliquity of the ecliptic (α) is exaggerated for illustrative purposes.

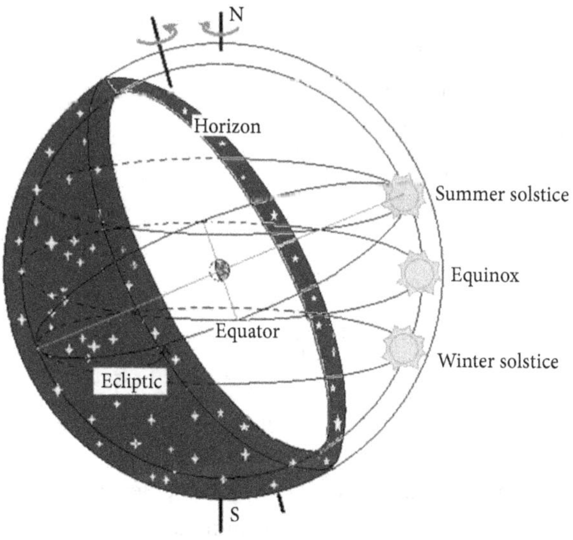

Fig. 6.3. The motion of the Sun as composed of two rotations, the diurnal rotation of the celestial sphere and the annual rotation of the Sun's own sphere.

Eudoxus attempted to account also for the broad features of the motions of the Moon and the planets by adding a number of spheres nested inside one another and rotating around different axes.

Astronomical theories derived from the two-sphere model, in which the Earth is at rest at the centre of the rotating cosmos, are called "geocentric" or "geostationary" (from *gaia*, the ancient Greek word for Earth.) The vast majority of Greek thinkers adopted theories of this kind, but a few dissenters were known. The most famous of these were Heraclides of Pontus and Aristarchus of Samos. Heraclides, a contemporary of Eudoxus, proposed to account for the diurnal cycle by assuming that the celestial sphere is stationary and the Earth rotates daily around its axis in the eastward direction. His idea was followed up fifty years later by Aristarchus, who argued not only that the Earth rotates around its axis, but it also revolves around the Sun, so as to explain the annual cycle. A theory like that of Aristarchus is called "heliocentric"—from *helios*, the Sun. The Greek astronomers knew that the astronomical phenomena of the diurnal and annual cycles provide no basis for deciding between a geocentric and a heliocentric theory. Nonetheless, they dismissed the notion that the Earth could be moving on the basis of terrestrial observations. It seemed entirely obvious that if there were any such motion, we who stand on the Earth would notice it. In addition to this conviction, deeply rooted in everyday experience, there was one astronomical argument that weighed against heliocentrism. If the Earth revolved around the Sun, our changing point of observation would cause the positions of the stars to appear different at different times of the year. This effect, which is called stellar parallax, would be very noticeable if the distance between the Earth and the stars were comparable to the Earth's displacement, and it becomes less evident the farther away the stars are (fig. 6.4). Since the Greek astronomers could observe no stellar parallax and could not fathom that stellar distances could be large enough to make the effect undetectable, they had one more reason to discard the heliocentric hypothesis.

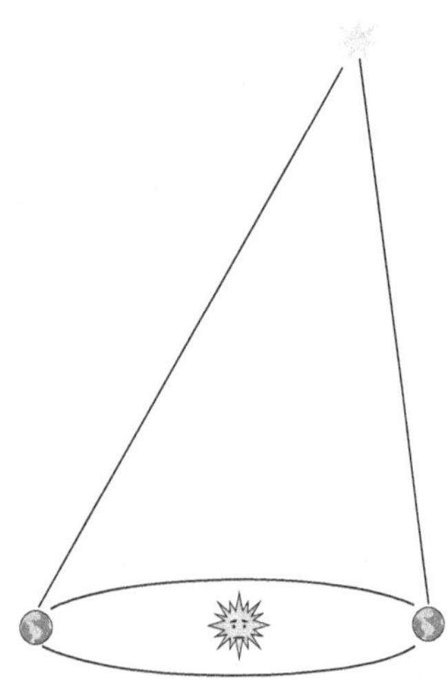

Fig. 6.4. The heliocentric hypothesis implied stellar parallax, but no stellar parallax could be detected with the naked eye. This was one reason why ancient Greek astronomers rejected heliocentrism.

A long-standing tradition attributes to Plato the belief that only uniform circular motion—a motion that is perfectly symmetrical and eternally equal to itself—is suitable to the perfect and unchanging nature of the heavens. Some nine centuries after Plato, Simplicius, in his sixth-century CE *Commentary on Aristotle's* On the Heavens, quoted first-century CE Sosigenes:

> Plato … set the mathematicians the following problem: What circular motions, uniform and perfectly regular, are to be admitted as hypotheses so that it might be possible to save the phenomena presented by the planets?

Although this story is probably apocryphal, the philosophical assumption that the motions of celestial bodies must be uniform and circular is traditionally known as "Plato's principle", and the task of devising a theory that incorporates it and at the same time explains the astronomical phenomena is known as "Plato's problem". Other late-Antiquity authors traced the principle of uniformity and circularity back to the Pythagoreans. What we know for sure is that by the time of Aristotle this principle was firmly established in Greek astronomy. Aristotle integrated it with his cosmology and his physics and, later, Islamic and Christian philosophers embraced it. The Aristotelian cosmos inherited by medieval and early modern Europeans was very similar to Eudoxus's model of geocentric spheres (fig. 6.5).

Whereas the belief in geocentrism was overwhelmingly prevalent but not absolute—other possibilities, as for example heliocentrism, were entertained at least hypothetically—the dogma of circularity and uniformity reigned unchallenged for two thousand years. It provided the theoretical foundation of every astronomical model that was developed

Fig. 6.5. A Christian rendition of the Aristotelian cosmos (Petrus Apianus *Cosmographia*, 1524).

© Heritage images/corbis.

during and after the Hellenistic Age, and would not be dislodged until the seventeenth century CE. What it meant to use mathematical "hypotheses" to "save the phenomena", however, was a question connected to the question of the relation between mathematics and the physical world. We will return to it after examining the development of mathematical astronomy in the Hellenistic period, as we know it from Ptolemy's *Almagest.*

6.3 Hellenistic Astronomy, Claudius Ptolemy, and the *Almagest*

The explanation of astronomical phenomena by means of uniformly rotating spheres became a fixture of Greek science. Astronomers treated it in the typical style of theoretical mathematics, with a set of first principles followed by theorems and demonstrations. It was also the subject of philosophical disquisitions, popular summaries for lay readers, and poetical descriptions. But if this theory was able to explain the broad features of the phenomena, the details presented challenges to the mathematical theorists. First of all, it was observed that the Sun presents an "anomaly" or "inequality" with respect to uniformity. If the Sun traveled with constant speed around a circle centered on Earth, the four seasons would have the same duration, while in fact they do not. The Greek astronomers measured that the time from spring equinox to summer solstice was about two days longer than that from summer solstice to fall equinox, and about six days longer than that from fall equinox to winter solstice. The Sun appears to travel a little faster in winter than in summer.[1] The planets, too, were seen to present inequalities analogous to that of the Sun. Furthermore, their retrograde motions violated both uniformity and circularity. Finally, their sizes appeared to change, as if their distance from the Earth did not remain constant.

By the late third century BCE, geometrical models more refined than those of Eudoxus were invented to incorporate these irregularities. One helpful simplification was obtained by ideally changing the point of view. Astronomers began representing the various motions as if they were observed from the celestial North Pole, so that the paths were pictured as lying flat on the plane of the ecliptic. In this way, the geometry involved was no longer three-dimensional but two-dimensional. Suitable corrections to the flat models were made to take into account the obliquities of the paths of the Moon and planets, which lie nearly but not exactly on the plane of the ecliptic.

[1] That is still the case, of course, but the specific lengths of the seasons have changed. Today summer is the longest.

Another significant change occurred in the third century BCE, as a result of the creation of the Hellenistic empire by Alexander the Great: the data of the Babylonian astronomers became available to the Greeks. The detailed Babylonian measurements and the Babylonian arithmetic methods to predict the positions of celestial bodies prompted on the part of the Greeks the realization that the existing theories were loosely explanatory but failed in the details. Greek astronomers matured a new conviction: it was not sufficient for a theory to offer satisfactory explanations of the qualitative features of the phenomena. A good theory had to reproduce the empirical observations and to enable the prediction of future observations in quantitative detail. In other words, the merging of Greek geometrical modeling with Babylonian measurements and numerical procedures gave rise to stricter criteria to judge whether an astronomical theory was or was not empirically adequate.

The first mathematician known to have worked out new mathematical models of astronomical phenomena was Apollonius of Perga, who lived in the second half of the third century BCE and studied in Alexandria of Egypt. No astronomical writing by Apollonius has survived, but Ptolemy informs us that Apollonius formulated and proved theorems about the models called "eccentric" and "epicycle on deferent" (see below.) Apollonius might have invented the two models as a solution to Plato's problem, but as far as we know he dealt with them purely from a geometrical point of view. The comparison between models and data was carried out by Hipparchus, in the second century BCE. Hipparchus integrated the old Babylonian measurements with new ones that he carried out in Alexandria, on the island of Rhodes, and possibly in his hometown of Nicaea in Asia Minor. Hipparchus also developed trigonometry, a new branch of mathematics that was necessary to use Apollonius's models, and formulated a theory of the Sun's motion that was later adopted by Ptolemy. Ptolemy built upon Apollonius's and Hipparchus's results to produce an astronomical theory that was both explanatory in the terms of the physics and philosophy of the time, and empirically adequate within the limits of the available observations.

Claudius Ptolemy's treatise known as *Almagest* is the most important ancient astronomical work that has arrived to us. Ptolemy's last name suggests that he was from one of the Egyptian cities named after the kings of the Ptolemy dynasty. He worked at the Museum of Alexandria and wrote a number of books in mathematics, astronomy, optics, harmonics, geography, as well as astrology. His major work was titled *Mathematike syntaxis* ("Mathematical Collection"), and became known also as *Megale syntaxis* ("Large Collection"). Islamic scholars translated it into Arabic in the ninth century; they called it *al-mjsty*, which probably derives from the Greek "magiste" ("greatest"). The most influential Latin version of the book was made from the Arabic in the twelfth century CE. Thus the book became known in medieval and early modern Europe by the Arabic title, Latinised into *Almagest.*

The Almagest occupies in the history of astronomy a place analogous to that occupied in the history of mathematics by Euclid's *Elements.* It is a compilation of the astronomical knowledge produced by several generations of Greek mathematicians, systematically organized and expanded by the author. It contains a stars catalogue, mathematical models for the Sun, the Moon, and the five planets as well as the mathematical tools (numerical tables, explanations, and proofs) necessary to understand the models. It is written in the impersonal style that has characterized theoretical mathematics ever since Euclid. In the first chapter, however, Ptolemy provided an overview of the philosophy behind his work. He classified astronomy as a branch of mathematics together with geometry and arithmetic, and mathematics as a branch of theoretical philosophy together with physics and theology. This classification coincided with that of Aristotle, except that Ptolemy placed astronomy completely within the realm of mathematics. This is not to say that he did not consider the celestial bodies to be material bodies, or was not interested in their physical nature. He dealt with the physical structure of the cosmos in another book titled *Planetary Hypotheses,* in which he tried to reconcile his two-dimensional models with the spherical cosmos of Aristotelian philosophy. He also wrote a massive treatise titled *Tetrabyblos,* where he expounded what he believed to be the astrological powers deriving from the physical nature of celestial bodies. In the *Mathematical Collection,* however, he dealt exclusively with the mathematical aspects of celestial motions. He made use of the models first proposed by Apollonius, the eccentric and the epicycle on deferent, and supplemented them with a third model of his own invention, the equant.

The Eccentric Model

One way of solving to the problem of solar without abandoning the principle of uniformity and circularity was to make the hypothesis that the Sun travels uniformly on a circular orbit that is not centered on Earth but on a point a bit removed from it (fig. 6.6). In this model, the distance of the Sun (S) from Earth (E) is not constant. The point of maximum distance is called apogee, the point of minimum distance is called perigee, and the center of the orbit is called eccenter (C). Since the Sun travels uniformly on the eccentric circle, the radius from C to the Sun sweeps equal angles in equal times. The line of sight from Earth, however, spans larger angles when the Sun is near the apogee than when it is

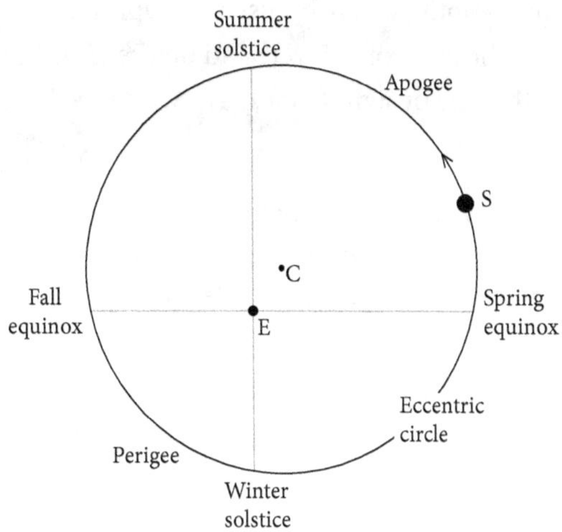

Fig. 6.6. The eccentric model.

near the perigee. From Earth the Sun appears to have a non-uniform angular speed and to move faster when it is closer to Earth. If the eccenter is positioned strategically, the model can reproduce the different lengths of the seasons. For the Sun, Ptolemy calculated that the eccentricity, that is, the distance of the eccenter from Earth, had to be equal to 1/24 of the radius of the eccentric circle.

The Epicycle on Deferent Model

Another model was used in order to reproduce the retrograde motions of the planets. In the simplest form, the model can be described as follows. A planet (P) revolves uniformly at a given speed on a circle called the epicycle, the center of which revolves uniformly at its own speed around the Earth (fig. 6.7).

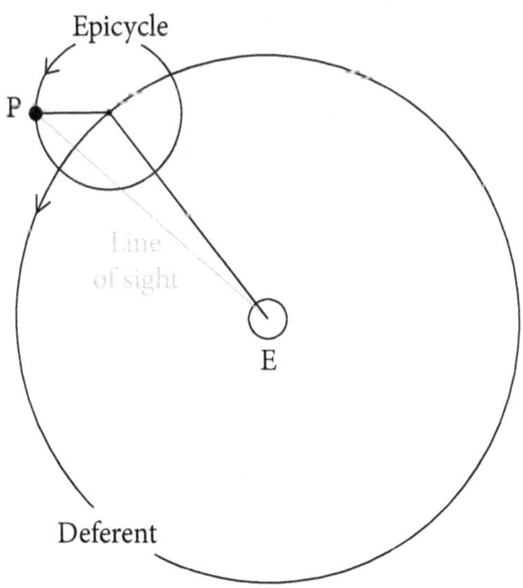

Fig. 6.7. The epicycle on deferent model.

The composition of the two uniform circular motions can be used to reproduce the observed motion of the planet. If the ratios of the radii of epicycle and deferent and the ratios of the two speeds are suitably chosen, the line of sight from Earth will trace loops that can reproduce the retrograde motions (fig. 6.8).

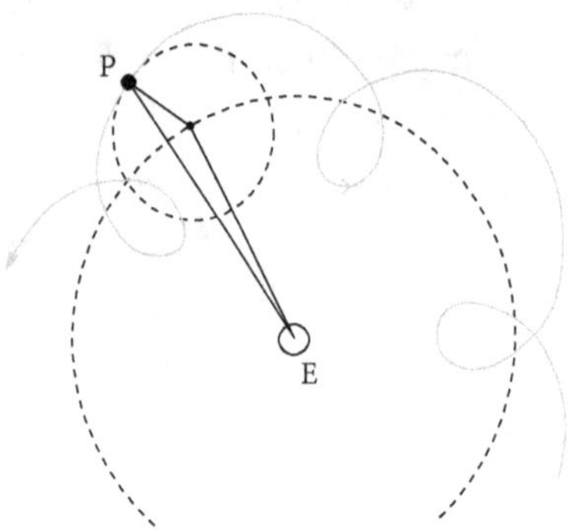

Fig. 6.8. Retrograde motion as a composition of two uniform circular motions in an epicycle on deferent model.

Ptolemy reported that Apollonius proved theorems about the resulting motion of the planet in an epicycle and deferent model, thus giving the conditions under which loops of different number and size are produced. He also showed that if the two component rotations had equal angular speeds, the epicycle on deferent model was mathematically equivalent to the eccentric model, for the resulting motion of the planet was uniform and circular, but around an eccenter that was removed from Earth by a distance equal to the radius of the epicycle (fig. 6.9). Thus, the solar inequality could be modelled either with an eccentric or with an epicycle on deferent.

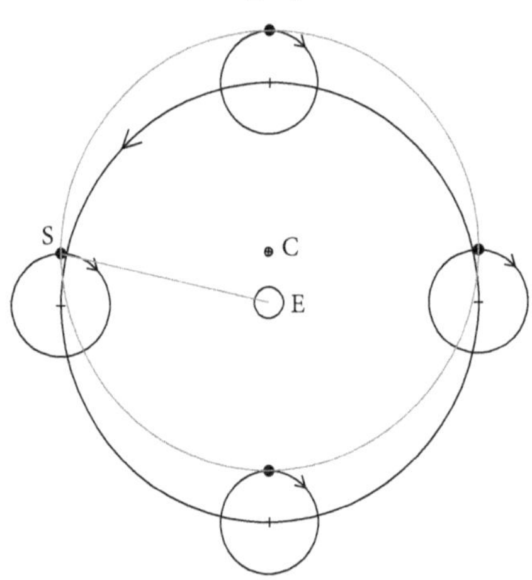

Fig. 6.9. Mathematical equivalence of eccentric model and epicycle on deferent model.

The Equant Model

In principle, the eccentric and epicycle on deferent models could be combined to reproduce both the retrograde motions and the inequalities of planets. Matching Apollonius's and Hipparchus's geometrical results to old and new data, however, Ptolemy found that planetary inequalities and retrogradations could be accurately reproduced only if the motion of the deferent was assumed to be uniform around a point separated from the centre of the deferent. He called this hypothetical point equant (Q), and calculated that the centre of the deferent, that is, the eccentER (C), had to be positioned on the mid-point between E and Q (fig. 6.10).

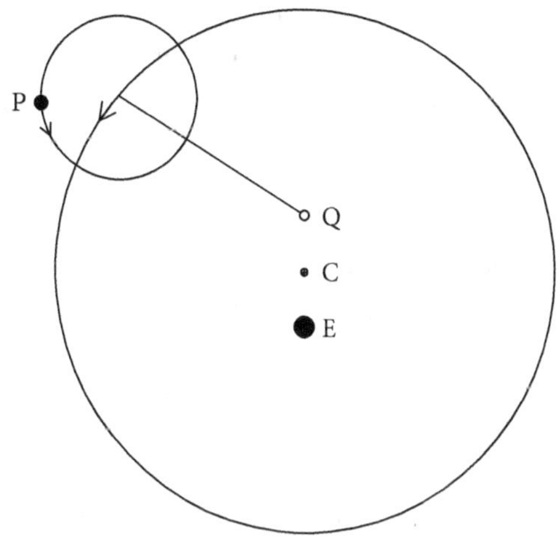

Fig. 6.10. Ptolemy's equant model.

6.4 Saving the Phenomena

Greek mathematical astronomy reached a high degree of empirical adequacy with Ptolemy's geometrical models of the Sun and the planets in *The Almagest.* It "saved the phenomena" within the limits of observational precision. Nonetheless, astronomical practice revealed that models that were physically different, as for example the eccentric model and the epicycle on deferent model for the Sun,were mathematically and empirically equivalent. It was simply not possible to decide between the two models from the standpoint of mathematical modeling and measurements alone. The question of which model reflected physical reality required considerations that went beyond mathematics and observations. Hipparchus, for instance, preferred the epicycle on deferent hypothesis because in his view it was more respectful of the centrality of the Earth. Ptolemy, in contrast, regarded the eccentric model to be more likely to represent reality because in his view it was physically simpler, as it involved one rotation instead of two. To Ptolemy, evidently, the belief in the simplicity of nature took precedence over the belief in the centrality of the Earth.

In the early part of the twentieth century, in the light of modern debates about the meaning of scientific theories, some philosophers of science interpreted the Greek efforts to "save the phenomena" in an instrumentalist sense. According to the instrumentalist view of the role of mathematics in science, a mathematical theory is simply an instrument useful to make calculations of physical quantities and to predict the outcome of measurements, without expecting the instrument to be able to also represent the phenomena realistically. A more complete reading of the ancient Greek astronomical literature, however, reveals that the Greek astronomers did expect their theories to be realistic representations of the physical world.

Ptolemy, for example, adapted the mathematical "hypotheses" of the *Almagest* to Aristotle's physical theory of the heavens. In a book titled *Planetary Hypotheses*, he argued that the heavenly bodies were contained in solid circular shells nested inside one another as they were in the Aristotelian cosmos.[2] The motions specific to each body were produced by the frictionless rotations of the shells around their axes (fig. 6.11).

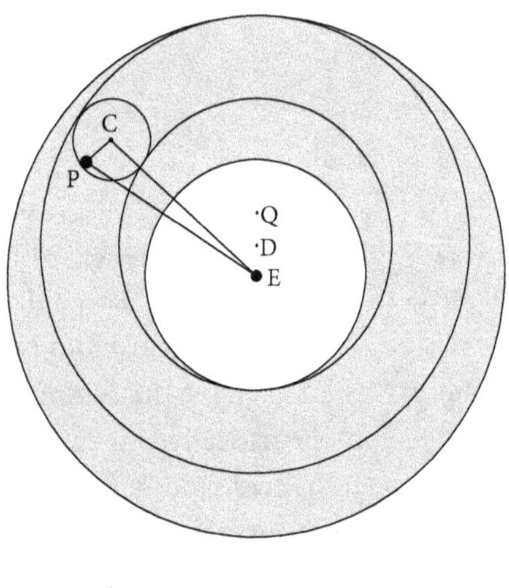

Fig. 6.11. A simplified illustration of the physical models of the planets in Ptolemy's *Planetary Hypotheses*. The shaded area represents the planet's sphere (more precisely, a spherical shell) as seen from the celestial pole. It is centered on Earth (E), around which it rotates daily, carrying an eccentric sphere centered on D (eccenter). The eccentric sphere rotates uniformly with respect to Q (equant), and it carries a smaller sphere, which rotates around its center (C) carrying the planet (P). The motion of C describes the eccentric deferent of the planet, while the motion of P around C describes the epicycle.

[2]See also Sect. 7.3

He thus integrated his mathematical hypotheses with Aristotle's physical structure, according to which the sphere of the Earth (sublunar sphere) was contained inside the sphere of the heavens (superlunar sphere), and the sphere of the heavens was subdivided into the concentric shells of the Moon, Mercury, Venus, the Sun, Mars, Jupiter, Saturn, and the fixed stars, with no empty space between one sphere and the next (see fig. 6.5). Thanks to this integration, Ptolemy was able to make an estimate of the sizes of the various spheres and of the whole cosmos. By his reckoning, the average distance of the Sun from Earth was 1210 times the radius of the Earth, and the inner radius of the eighth sphere, the sphere of the fixed stars, was 19865 times the radius of the Earth.

Despite their interest in physical matters, Greek mathematicians viewed mathematics as separate from physics. In particular, they required their mathematical models to be empirically adequate but did not infer physical explanations of the phenomena from the models. They were aware that different models could be mathematically equivalent and could therefore save the same phenomena to the same degree of empirical adequacy, even though if taken realistically they would represent very different physical mechanisms. Hence they kept mathematical modelling separate from the analysis of the nature of the physical world.

Sources

Cuomo, Serafina. *Ancient Mathematics.* London and New York: Routledge, 2001.

Evans, James. *The History and Practice of Ancient Astronomy.* Oxford: Oxford University Press, 1998.

Evans, James and Berggren, J. Lennart. *Geminos's Introduction to the Phenomena. A Translation and Survey of Astronomy.* Princeton: Princeton University Press, 2006.

Hussey, Edward. "Aristotle and Mathematics". In *Science and Mathematics in Ancient Greek Culture,* edited by C. J. Tuplin, T. E. Rihll, and L. Wolpert. Oxford Scholarship Online, January 2010; print publication date, 2002.

Katz, Victor J. *A History of Mathematics. An Introduction.* 3rd Edition. Boston: Addison-Wesley, 2009.

Lloyd, Geoffrey E. R. "What was mathematics in the ancient world?". In *The Oxford Handbook of the History of Mathematics,* edited by E. Robson and J. Stedall, 7–25. Oxford and New York: Oxford University Press, 2009.

Netz, Reviel. *The Shaping of Deduction in Greek Mathematics.* Cambridge: Cambridge University Press, 1999.

Pedersen, Olaf. *A Survey of the Almagest.* New York: Springer, 2010.

Saito, Ken. "Reading ancient Greek mathematics". In *The Oxford Handbook of the History of Mathematics,* edited by E. Robson and J. Stedall, 801–826. Oxford and New York: Oxford University Press, 2009.

Chapter 7

Mathematics and the Origin of Modern Science Part 1

7.1 The Copernican Revolution, the Scientific Revolution, and the Birth of Modern Science

Nicolaus Copernicus's treatise titled *De revolutionibus orbium coelestium* (in English, *On the Revolutions of the Heavenly Spheres*) was published in 1543. In it, the Polish astronomer presented a detailed mathematical demonstration that a *heliocentric* theory, in which the Sun is stationary as the Earth revolves around it, was as capable of "saving the phenomena" as well as the ancient *geocentric* theory expounded in Claudius Ptolemy's *Almagest* did. The scientific and cultural impact of this highly technical book, originally addressed to a restricted readership of mathematicians and astronomers, increased enormously in the seventeenth century among specialists as well as the general public. The eventual triumph of Copernicus's heliocentrism came to be seen as an epochal event that started a radical renewal of science—of the fundamental assumptions on which the scientific view of the world rests, as well as of the methods of scientific enquiry and therefore of the very nature of scientific knowledge. Modern science is widely regarded to be the result of this transformation, which is now known as the Scientific Revolution. Although the word "revolution", with its political connotations, evokes a single sudden event responsible for shattering a whole pre-existing regime, it is easy to recognize that the Scientific Revolution was a long and complex process involving a variety of factors and activities.[1] Even remaining within the field of astronomy, the transformation initiated by Copernicus required about 150 years and the work of many others—most famously of Galileo Galilei, Johannes Kepler, and Isaac Newton—to be in some sense completed. It is noteworthy that in the process certain fundamental assumptions, which Copernicus had inherited from his predecessors and had taken for granted, were discarded, and astronomy was changed far beyond what Copernicus had expected. For this reason, the historian of science Thomas Kuhn called Copernicus's *De revolutionibus* a "revolution-making" rather than a revolutionary book.

The expression "a Copernican revolution" is often used today to describe any radical change of point of view that generates a whole new way of understanding the world. This usage illustrates the high status achieved by Copernicus's work in our culture, but it also highlights a striking aspect of it. Simplistic conceptions of scientific progress rely on the idea that old theories are discarded when they are falsified by new observations. In the case of Copernicus, however, no new phenomenon had been discovered since the time of Ptolemy, and no new data were collected, which would have justified the abandonment of geocentrism and the shift to a heliocentric theory. The change that occurred was a change in the way in which the known phenomena were modelled. In its broader implications, it was also a deep change in the interpretation of the function of mathematical modeling and in the role assigned to mathematics in the scientific enterprise. Although many features of science have changed since the Scientific Revolution, the importance of mathematics to the scientific understanding of natural phenomena, forcefully asserted by Copernicus and his followers, has only increased.

[1] The history of the term "revolution" is interesting. Originally, it meant the circular motion of a celestial body revealed by cyclical astronomical phenomena, as is made clear by Copernicus's title, *On the Revolutions of the Heavenly Spheres.* In the sixteenth century the word began to be used to refer to political transformations that were meant to re-turn to or restore an earlier state of affairs, according to a cyclical conception of history. Over the course of the seventeenth and eighteenth century, a linear and progressive conception of history became prevalent, and the term started to be applied to political overturns that result from violent revolts, such as the French and American Revolutions. Although occasionally used also for scientific advancements by the philosophers of the Enlightenment and of Positivism, the word returned to astronomy in the twentieth century, when the expressions "Copernican Revolution" and "Scientific Revolution" gained currency.

7.2 The Place of Mathematics in the Early Universities

Cultural life in Europe between the thirteenth and seventeenth centuries revolved mainly around universities. Although demands of practical training for commerce, banking, navigation, and engineering by the emerging middle classes were on the rise—as is shown by the growth of abacus schools and the development of calculation techniques and algebra—the traditional academic education remained dominant. In universities, both the content and the organization of the teaching were based on Aristotelian philosophy, which had been purged of its pagan elements and integrated with Christian doctrine. The result was Scholastic Aristotelianism, or Scholasticism, a system of thought that conformed to the ideology of the Church. In this system, philosophy was considered the "handmaiden" of theology. In particular, natural philosophy, that is, the branch of philosophy that studied the natural world, was the study of God's creation, and the purpose of it was to learn about God.

Pre-industrial universities were not organized according to the specialized disciplines that we have today. The organization of learning was hierarchic. The standard university curriculum consisted of a preliminary course of studies in the faculty of arts, leading to the degree of "Master of Arts". The students could then move on to higher studies in the faculties of law, medicine, and theology, at the end of which they gained the title of "Doctor". The faculty of arts taught the seven liberal arts of the Trivium, consisting of grammar, logic, and rhetoric, and of the Quadrivium, consisting of arithmetic, geometry, astronomy, and harmonics. In other words, mathematics and the two physical-mathematical sciences of astronomy and harmonics were taught as introductory general subjects in preparation for the more advanced and specialized disciplines. While arithmetic and geometry were concerned with abstract entities, astronomy and harmonics were classified as "mixed mathematics" or "mixed sciences", as they were concerned with the material world but studied it mathematically. There existed a spectrum of views on the power of mathematical methods in understanding nature, but the prevailing view was that mixed mathematics was not part of natural philosophy because it lacked the ability to uncover the causes of phenomena. Mathematical models could describe the results of observations, but only philosophy could reveal *why* things occurred the way they did. The task of a mathematician was limited to modelling and reckoning the results of past and future observations, while it was the job of a philosopher to explain the real nature of the objects observed. Most academics held an *instrumentalist* view of the relationship between mathematics and physical reality. They considered the models or hypotheses of mixed mathematics to be just instruments for calculation and prediction, not representations of reality. Astronomy, in particular, was valued for two reasons. On the one hand, it prepared the mind for the contemplation of the beauty and perfection of God's creation. On the other hand, it was useful for calculating the dates of religious celebrations, reckoning the times of biblical chronologies and prophecies, assisting in navigation, architecture, and engineering, and making horoscopes and astrological prognostications. Nonetheless, the hierarchy of academic disciplines was mirrored in the social hierarchy of the academic professions, which placed theologians at the top of the scale of prestige, intellectual authority, and financial rewards, and confined mathematicians to the most junior, subordinate, and correspondingly under-paid ranks.

Starting in the fifteenth century in Italy, amidst the broad social and economic changes brought about by the end of feudalism, the intellectual movement of Humanism led the way to the European Renaissance. The humanists advocated a re-birth of culture (*renaissance* is the French word for re-birth), to be achieved by a return to the ideas and forms of expression of the classical Greek and Roman era, after what they saw as the barbarism of the "Middle Age", that is, the centuries that followed the fall of the Roman Empire. While Humanism was mainly focused on language, literature, art, and moral and political thought, it also involved a search for the original texts of classical mathematics and astronomy. Thus, the Renaissance involved, alongside the renewal of literature, poetry, architecture, and art, also a "mathematical Renaissance". In various places and in various forms, humanistic scholars advocated a larger role for mathematics and the mathematical sciences in the education of the ideal citizen. They lamented the neglect of the mathematical disciplines in universities and the subordinate role reserved to them by the Scholastic mindset. They promoted instead the intellectual value of mathematics, stressing its logical rigour and its ability to attain stable and certain results in contrast to the vagaries and uncertainties of philosophy. The mathematical Renaissance began as an ambition to return to the (real or imaginary) splendours of classical Antiquity, the memory of which was presumed to have been lost or corrupted during the Middle Age. Thus, for example, François Viète and, initially, René Descartes aimed at re-discovering "the lost art of analysis".[2] However, as the study of the ancient books progressed and the number of competent readers increased (the birth of the typographical industry and of

[2]See Sect. 2.2.

the book market played no little part in this), a new kind of mathematicians emerged. They no longer viewed the old knowledge as unsurpassable, but were confident in their own ability to create a new science, building upon the most valuable parts of the classical scientific legacy in order to move forward. The courts of the Renaissance princes, where mathematicians were offered employment and patronage in exchange for their contribution in giving lustre to and increasing the military power of the prince, constituted new venues where mathematical knowledge could grow outside the rigid structure of the universities, and created opportunities for the new science to emerge. Also, new educational institutions began to be founded in the latter part of the sixteenth century. These included the colleges of the order of Jesuits, where the teaching of mathematics was a strong part of the curriculum, as well as new schools for military and naval engineers, where mathematics began to be used to advance technology.

7.3 The Christian-Ptolemaic-Aristotelian Cosmos

According to Aristotelian philosophy, the material world was a finite and spherical *cosmos*, or ordered whole, made of interconnected parts. The Earth was spherical and stationary and it sat at the centre of the cosmos. There existed no void—the cosmos was everywhere full of matter. The cosmos was divided into two main regions, the terrestrial sphere, containing the Earth and the surrounding atmosphere, and the celestial sphere, which was, more accurately, a spherical shell surrounding the terrestrial sphere and containing the celestial bodies. Everything in the terrestrial sphere was made of four elements, earth, water, air, and fire. Each element had its proper natural place, to which it naturally returned when free to move according to its nature. The natural place of earth was the centre of the cosmos. For this reason, bodies that were made prevalently of earth tended naturally to fall downward. The Earth was spherical and at rest because every part of it tended toward the centre. The natural place of water was the spherical shell surrounding the sphere of earth, that of air the spherical shell surrounding water, and that of fire the spherical shell surrounding air. That explained why water fell down when it was in air and oceans surrounded the Earth, and why airy substances and flames rose upward from the surface of the Earth. Natural motions (as opposed to forced or violent motions) in the terrestrial sphere were rectilinear, either toward or away from the centre. In contrast, the celestial region was made of a fifth element called "aether" (or "ether"), which was solid but completely transparent and frictionless. The natural motion of aether was uniform rotation around the centre. The various celestial bodies had their natural places in concentric spherical shells, with the innermost shell occupied by the moon and the outermost by the fixed stars (see fig. 6.5).[3] For this reason, the celestial region was also called the "superlunar sphere" (the sphere above the Moon, *luna* in Latin) and the terrestrial region the "sublunar sphere" (the sphere below the Moon). In Aristotelian natural philosophy, superlunar physics was fundamentally different from sublunar physics.

The master text of mathematical astronomy was the Latin translation of Ptolemy's *Almagest*.[4] In this technical treatise, Ptolemy expounded his models for "saving the phenomena" using simple or compounded circular paths for the Sun, the Moon, and the planets. The textbooks used in universities were simplified versions of the *Almagest*. They taught the basics of Ptolemy's models and some of the techniques to calculate the positions of the heavenly bodies, leaving out the full mathematical apparatus. The technical level of university astronomy was in general considerably lower than that achieved by Ptolemy, as it was considered sufficient to the practical demands of calendar computations and astrology. For example, horoscopes typically required the location of the Sun, Moon, and planets at a certain date within one of the zodiacal signs—Aquarius, Leo, Libra and so on—and for this it was sufficient to find their position within a 15° sector. Material models, in bronze, wood, and other materials, called armillary spheres were used to demonstrate the simplified mathematical models and to aid in the calculations. Sometimes they were provided with mechanisms that simulated the celestial motions, thus functioning as astronomical clocks. They were often feats of artisanal and technical skills and, at the same time, beautiful and expensive objects of art. In a sense, they embodied materially what mathematical models were in abstract form. They were useful artifacts and objects of admiration, which depended on philosophical principles for their fundamental blueprint. They were analogous to the phenomena that they simulated but had no pretense of revealing, by themselves, the true nature of the natural entities and processes.

In his book titled *Planetary Hypotheses*, Ptolemy had moved beyond the mathematical aspects of astronomy in an attempt to integrate the circular paths with the spheres of

[3] The addition of a ninth sphere (also called *crystallinum* in medieval astronomy) was to allow the sphere of the fixed stars to have a motion compounded of two rotations, the daily rotation and a slow rotation around the north-south axis. The second rotation accounted for the phenomenon, already observed by Hipparchus and taken into account by Ptolemy, called the precession of the equinoxes. Ptolemy attributed to the sphere of the stars a speed of about 1° in 100 years. A more accurate value would have been about 1° in 70 years.

[4] See Chap. 6.

Aristotle's physical theory.[5] He imagined the heavenly bodies to be encased in solid circular shells that fit inside one another and rotated without any friction or resistance around their respective axes, in such a way as to approximate the structure of the Aristotelian cosmos while at the same time allowing for the motions specific to each body (fig. 7.1).

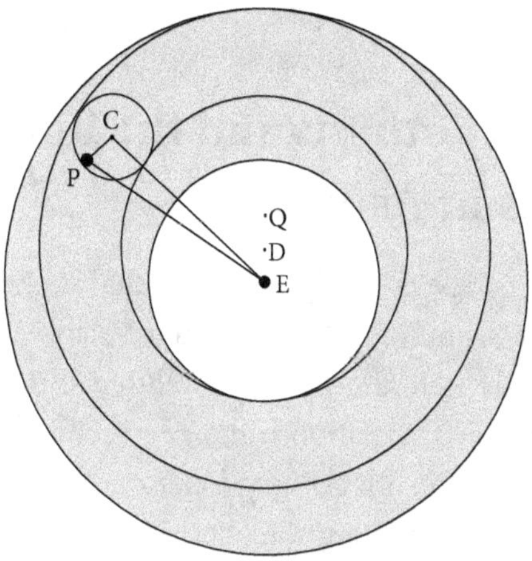

Fig. 7.1. A simplified illustration of the physical model of planetary motion according to Ptolemy's *Planetary Hypotheses*. The shaded area represents the planet's sphere (more precisely, a spherical shell) as seen from the celestial pole. It is centered on Earth (E), around which it rotates daily, carrying an eccentric sphere centered on D (eccenter). The eccentric sphere rotates uniformly with respect to Q (equant), and it carries a smaller sphere, which rotates around its center (C) carrying the planet (P). The motion of C describes the eccentric deferent of the planet, while the motion of P around C describes the epicycle.

Ptolemy adopted the traditional ordering of the celestial bodies, with the Moon as the closest, followed by Mercury and Venus, the Sun, Mars, Jupiter, Saturn, and the stars. While he had reasons for placing the Moon closest to Earth and the Sun above the Moon, Mercury, and Venus, he admitted that the ordering of the inferior planets (Mercury and Venus) and that of the superior planets (Mars, Jupiter, and Saturn) was arbitrary. He estimated the minimum and maximum distances of the Moon from Earth to be equal respectively to 33 and 64 times the radius of Earth. Hence, the radius of the sublunar sphere was 33 times the radius of Earth. He also estimated the minimum and maximum distances of the Sun from Earth to be equal to 1160 and 1260 times the radius of Earth.[6] From these values, making the assumption that the spheres of the celestial bodies fitted into one another as tightly as possible, he estimated the innermost and outermost radii of the spheres of the planets. The outermost radius of the sphere of the most distant planet,

[5]See also Sect. 6.4.

[6]Ptolemy's mean distance for the Moon is close to the modern value of the mean radius of the Moon's orbit, but his mean distance for the Sun is about 19 times too small.

Saturn, which coincided with the inner radius of the sphere of the fixed stars, turned out in Ptolemy's calculation to be equal to about 20000 times the Earth radius, which is less than the actual Earth-Sun distance. The part of *Planetary Hypotheses* in which Ptolemy estimated the size of the cosmos was well known and influential in medieval and Renaissance universities. Viewed from a modern perspective, the cosmos of the Aristotelian-Ptolemaic-Christian tradition was a round, full, well-ordered, and pretty small place, thoroughly interconnected by visible and invisible influences.

7.4 Copernicus's Heliocentric Hypothesis and the Role of Mathematical Hypotheses in Astronomy

Nicolaus Copernicus was educated at the universities of Krakow and Padua at the height of Humanism. One of the ancient texts that the humanists wanted to re-discover in pristine form was the *Almagest.* Although already well known from twelfth century translations, Ptolemy's treatise was the object of a new translation and a new technical analysis by the most accomplished astronomer of the fifteenth century, Johannes Müller of Königsberg, known by the Latinized name of Regiomontanus. Copernicus studied Regiomontanus's summary and critique. He was thus thoroughly trained in Greek mathematical astronomy, but followed Regiomontanus's assessment that the discipline was in need of reform. He found Ptolemy's results unsatisfactory and sought a way to improve upon them. In humanistic fashion, he did not present his own theory as a novelty, but rather as a return to pre-Scholastic ideas:

> I undertook the task of rereading all the works of all the philosophers which I could obtain to learn whether anyone had proposed other motions of the universe's spheres than those expounded by the teachers of mathematics in the schools. And in fact I found in Cicero that Hicetas [i.e., a Pythagorean] supposed the Earth to move. Later, I also discovered in Plutarch that certain others were of this opinion. [...] Therefore, having obtained the opportunity from these sources, I too began to consider the mobility of the Earth. And even thought the idea seemed absurd, nevertheless I knew that others before me had been granted the freedom to imagine any circles whatever for the purpose of explaining the heavenly phenomena. (Copernicus, 1992, "Preface")

Copernicus proposed to account for the celestial phenomena by replacing the assumption of a central and stationary Earth with the assumption that the Sun was stationary and the centre of the spheres of the stars and planets, as well as of the sphere of the Earth, and that the Earth was the centre of the sphere of the Moon. In this picture, the diurnal cycle is a phenomenon created by the rotation of the Earth around its axis, and the annual cycle and the retrograde motion of the planets are created by the revolution of the Earth's sphere around the Sun. (Copernicus also assumed a third motion, a conical wobble of the Earth's axis, to account for the precession of the equinoxes) (fig. 7.2). He then demonstrated that these hypotheses were mathematically and empirically equivalent to those of Ptolemy. From the point of view of "saving the phenomena", Copernicus's heliocentric theory worked as well as the old geocentric theory.

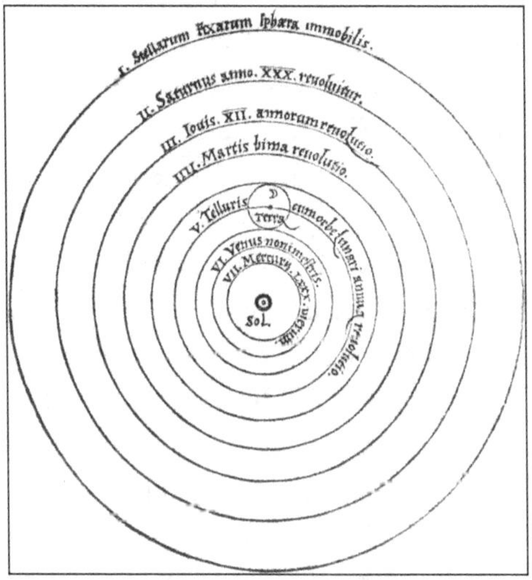

Fig. 7.2. Copernicus's heliocentric model.

What exactly convinced Copernicus to reject the millenary legacy of geocentrism in favour of a heliocentric theory is a matter of debate among historians. A few reasons, however, were offered by Copernicus in his introduction to the book. Firstly, he noted that astronomers disagreed with one another on the models to be used and could not even determine an exact standard length of the solar year. Secondly, he noted that in order to reach an agreement with observations, they were compelled to use Ptolemy's equant, which he saw as a violation of the ancient principle of uniform circular motion. In his interpretation, this principle meant that the motion of a heavenly sphere had to be uniform around the centre of the sphere—a condition unfulfilled by Ptolemy's models. Finally, he complained that traditional astronomers neglected the "structure of the universe and the symmetry of its parts." He compared them to "someone taking from various places hands, feet, a head, and other pieces, very well depicted, maybe, but not for the representation of a single person; since these fragments would not belong to one another at all, a monster rather than a man would be put together from them." (Copernicus, 1992, "Preface".) In contrast, Copernicus argued, his theory was not only more coherent with the principle of uniformity, but also more symmetrical and harmonious.

Copernicus achieved the same degree of empirical accuracy as the best geocentric theory, but without using any equant—an accomplishment that his followers highlighted and praised. Furthermore, Copernicus was able to account for the retrograde motions without the use of epicycles, by showing how the appearance of reversals in the motions of a superior planet resulted simply from the fact that we observe them from a viewpoint that also revolves around the same centre (fig. 7.3). He was unable, however, to avoid completely the use of epicycles and eccentrics, as he needed them to account for the inequalities of the planetary motions (epicycles used for this purpose are called minor epicycles). This meant that, aside from the avoidance of the equant, Copernicus's theory was equivalent to Ptolemy's also in terms of mathematical simplicity, if mathematical simplicity is evaluated in terms of the number of epicycles. Copernicus argued that his theory was preferable to Ptolemy's because of its overall symmetry and accord between its parts. While Ptolemy's models described the motions of individual planets adequately, they were unable to account for the connections that were observed between the planets and the Sun. The inferior planets, in fact, are always observed within a fixed angular distance (in technical language, they have a maximum elongation) from the Sun, like "dogs on a leash", and never in opposition. The superior planets can be at any angular distance from the Sun (they have no maximum elongation), but they are seen to retrograde only when they are in opposition. Ptolemy had taken care of these effects by means of convenient additional hypotheses. For the inferior planets, he had assumed that the centres of the epicycles of the inferior planets were always aligned with the straight line joining the Earth to the Sun, and for the superior planets he had assumed that the radius of the epicycle remained always parallel to the line joining Earth and Sun. In contrast, the heliocentric theory explained the observed connections between Sun and planets simply by virtue of the revolutions around a common centre, without the need for any *ad hoc* hypothesis (figs. 7.3 and 7.4).

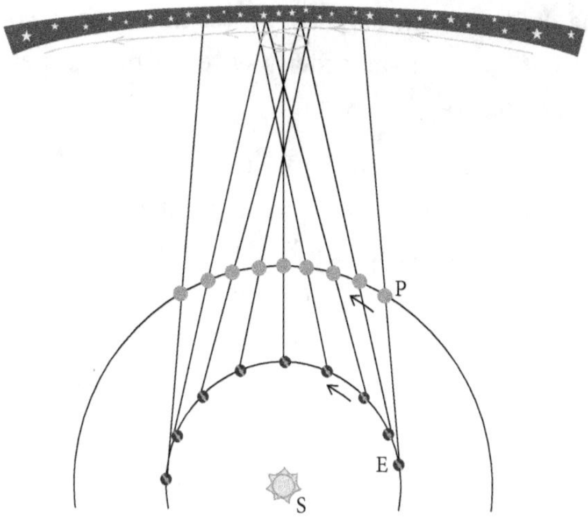

Fig. 7.3. In the heliocentric model, a superior planet (P) appears to have retrograde motion as an effect of the revolution of both the planet and Earth (E) around the Sun (S). This also explains why retrogradations are observed only when the planet is in opposition to the Sun.

Mathematicians counted this economy of hypotheses as an advantage, since it was a merit for a mathematical theory to derive the largest possible number of theorems from a minimum number of postulates.

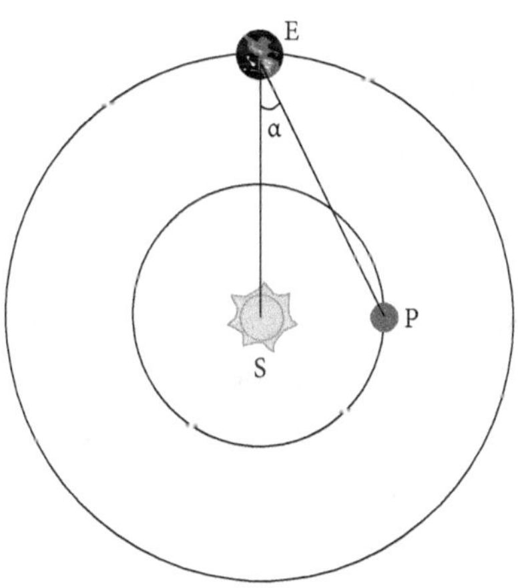

Fig. 7.4. In the heliocentric model, an inferior planet is always within a maximum elongation (α) from the Sun.

Finally—and probably this was for Copernicus the strongest advantage—the heliocentric hypothesis eliminates any arbitrariness in the definition of the order and relative sizes of the planets. Once the assumption is made that the Earth and planets revolve around the Sun, the overall structure of the whole system can be completely determined on the basis of the observed phenomena. Better to say, the theory gives the ensemble of heavenly bodies a definite structure, uniting them into an orderly system. Thus, Copernicus praised the order, symmetry, and harmony of his theory:

> Having thus assumed the motion which I ascribe to the Earth, by long and intense study I found that if the motions of the planets are correlated with the orbiting of the Earth, not only do their phenomena follow therefrom, but also the order and size of all the planets and spheres, and heaven itself is so linked together that in no portion of it anything can be shifted without disrupting the remaining parts and the universe as a whole. (Slightly adapted from Copernicus, 1992, "Preface".)

He repeated this point later in the book, adding—in the eloquent style that was typical of humanist rhetoric—that it was only appropriate to place the Sun at the center of the "family of planets", as it is the "lamp" that gives light to the whole: [7]

> At rest in the middle of everything is the Sun. For in this most beautiful temple, who would place this lamp in another or better position than that from which it can light up the whole thing at the same time? [...]Thus indeed, as though seated on a royal throne, the Sun governs the family of planets revolving around it. [...]In this arrangement, therefore, we discover a marvelous symmetry of the universe, and an established harmonious linkage between the motion of the spheres and their size, such as can be found in no other way. (Slightly adapted from Copernicus 1992, Book 1, Chapter 9.)

In Copernicus's system, the distance of Mercury from the Sun was approximately 0.4 of the Earth-Sun distance, that of Venus 0.7, that of Mars 1.5, that of Jupiter 5, and that of Saturn 9. (The exact values that can be calculated from his scale parameters all differ less than 3% from the modern values.) In order to explain the absence of any observable stellar parallax—which since antiquity had been one of the major objections to the hypothesis of an annual motion of the Earth—Copernicus proposed that the sphere of the stars was sufficiently large to make the Earth's orbit immeasurably small by comparison. In his system, the stars were vastly more distant than Ptolemy had estimated.

[7]Neoplatonism, a philosophy fashionable among humanists, assigned to the Sun a special place in the cosmos, as the source from which light and life emanated. See also next section.

De revolutionibus orbium coelestium appeared in print in 1543, a few months after Copernicus's death. The scholar who took care of the publication, a Lutheran theologian called Andreas Osiander, felt the need to add a defensive (and unsigned) "Letter to the Reader" to the book. He argued that Copernicus's heliocentric theory was to be taken into consideration not as "true or even probable", but as a mathematical device, "a calculus consistent with the observations." The task of a mathematician, Osiander observed, was to provide useful computations, not to speculate about the causes of phenomena. No one should expect astronomy to deliver certainty about hypotheses, "lest he accepts as the truth ideas conceived for another purpose, and depart from this study a greater fool than when he entered it." (Copernicus 1992, "Foreword by Andreas Osiander".) Osiander's argument was in line with the traditional instrumentalist interpretation of mathematical models, and it helped to get the book accepted, in instrumentalist terms, by astronomers.

Copernicus did not have the opportunity to comment on the added foreword, but he probably would not have approved of it. He had made clear that he meant his theory to be a realistic representation of the physical cosmos. This is shown not only explicitly by his words but also implicitly by his depiction of the cosmic spheres in the style of the traditional Scholastic-Aristotelian heavenly spheres (fig. 7.2). It was also demonstrated by his rejection of the equant because it could not be reconciled with the physics of uniformly rotating solid spheres. Copernicus was obviously aware that the idea of a moving Earth appeared indefensible, and had been treated as such by all the astronomical authorities since Antiquity. He attempted to suggest a possible modification of Aristotelian physics that could "legitimize the absurd" (to use the expression of the historian Robert Westman) and reconcile the Earth's motion with the facts of daily experience. He argued that the "center of gravity", that is, the centre toward which heavy objects naturally moved and hence the centre of the Earth, did not need to coincide with the centre of the cosmos. He further posited that the uniform rotational motions that he attributed to the Earth were natural motions of all the terrestrial elements. Therefore, every object in the terrestrial sphere would naturally participate in the Earth's rotations, making them undetectable by us. Copernicus was probably aware that this argument was too vague to meet the standards of natural philosophy. Nevertheless, he was confident enough in his theory to uphold it not only as a mathematical hypothesis but also as a physical theory. His confidence derived from the mathematical merits of the theory, even in the face of what seemed physical absurdity.

Anticipating the objections against geocentrism, he did not retreat into an instrumentalist position as Osiander did. On the contrary, he spurned the "babblers who are ignorant of astronomy" and proudly declared, "Mathematics is written for mathematicians." (Copernicus, 1992, "Preface".)

It is not surprising that Osiander felt Copernicus's book needed a defense. Not only did heliocentrism appear to clash with common sense but also with some passages in the Bible that asserted that the Earth was at rest and the Sun rotated around it—as some readers were quick to point out. Nevertheless, he also knew that it was inappropriate for an astronomer to propose a new physical theory of the heavens, for it transgressed the accepted hierarchy of academic disciplines, which placed physics on a superior level than the sciences of the Quadrivium. Osiander's strategy proved successful, for mathematical astronomers were left free to adopt Copernicus's theory as a mathematical hypothesis without having to commit to its physical reality.

In the view of the majority, the value of mathematical hypotheses lay in their usefulness. The Roman Church was at the time in the process of producing a much needed calendar reform. The old Julian calendar, which had been established by Julius Caesar in 45 BCE, was based on a four-year cycle in which three years of 365 days were followed by an intercalation year, or leap year, of 366 days. Thus, it defined an average year of 365 ¼ days. This value, however, is a few minutes longer than the actual time from one spring equinox to the next spring equinox. The small discrepancy had accumulated over the centuries and by the sixteenth century it had caused the conventional date of the spring equinox, March 20th or 21st, to shift ahead of the astronomical equinox, which now fell around March 11th. The exact determination of the equinox time was a problem for the Church, as the date of Easter depended on it. Reportedly, Copernicus had been consulted on the problem of establishing a new calendar, and he had replied that a reform was premature since astronomers were still unable to measure the exact length of the year with the necessary precision. Thanks to the fact that mathematics in general, and the *Almagest* in particular, were made part of the humanist revival of classical learning, when Copernicus's book became available, there were at least a few astronomers who were competent enough to appreciate the technical power of the new theory. They used it to calculate new astronomical tables, which were then used by Church officials to produce a new calendar. The reformed calendar, called Gregorian after the pope who issued it, Gregory XIII, was made official in Catholic countries in 1582. It was eventually adopted also in northern Europe and, with some later refinements, it is the calendar that we still use today.

7.5 Kepler and the Demise of the Celestial Spheres

One of the most influential aspects of Humanism was Renaissance Neoplatonism, a revival in a Christian key of Neoplatonism, which was the late-Antiquity interpretation of Plato's philosophy. For our purposes, the relevant parts of Renaissance Neoplatonism were a return to the Platonic belief that physical reality is shaped upon mathematical forms, and a quasi-religious cult of the Sun as the source of all power and life in the universe. A fervent supporter of these Neoplatonic ideas was Johannes Kepler, a mathematician who began his career teaching at the University of Graz in Austria. Kepler embraced Copernicus's theory; he also believed that mathematicians should allow themselves to be concerned with physics (and thereby philosophy) even more than Copernicus did. As he wrote in the preface of his first book,

> While Copernicus had attributed motion to the Earth rather than to the Sun on the basis of mathematical reasons, I did it on the basis of physical, or better still, metaphysical reasons. (Kepler, *Mysterium Cosmographicum*, Preface.)

In this book, which he titled suggestively *Mysterium Cosmographicum* ("Cosmographical Mystery"), he sought to explain the number, order, and relative sizes of the Copernican spheres by means of the geometrical properties of the five regular polyhedra, also known as the five Platonic solids. The discovery that there exist only five polyhedra having equal regular polygons as faces had been made in ancient Greece. This was probably around the time of Plato's founding of the Academy, and had inspired Plato to imagine that the physical world was created by a "Divine Craftsman", or "Demiurge", out of these geometrical objects. Plato had proposed an evocative, if puzzling, "geometrical atomism", according to which the particles of the four elements were shaped like the solids, earth particles like cubes, water particles like icosahedra, air particles like octahedra, and fire particles like tetrahedra, while the cosmos as a whole was modeled as a dodecahedron (fig. 7.5).

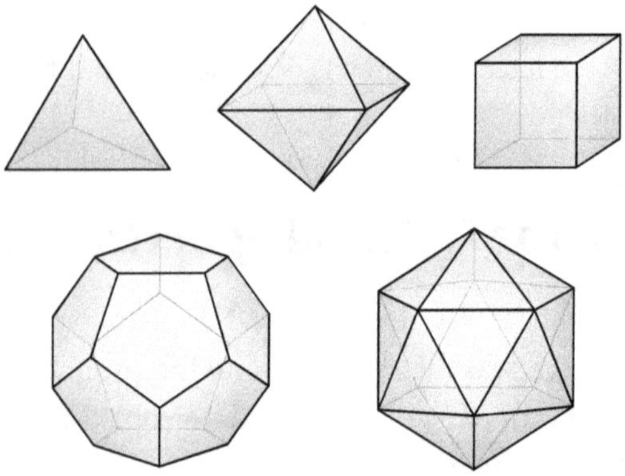

Fig. 7.5. The Platonic solids. Clockwise from top left: tetrahedron, octahedron, hexahedron, icosahedron, dodecahedron.

Like Plato, Kepler envisioned a role for these symmetrical shapes in the constitution of the world, but instead of seeking them in the minute structure of matter, he imagined that they could "explain" the planetary spheres. If six concentric spheres are nested inside one another in such a way that the five polyhedra are tightly interspersed with them (more precisely, circumscribed on and inscribed in them), the ratios of the radii of the spheres are geometrically determined. Kepler noticed that if he chose the order of the polyhedra suitably, the resulting ratios were surprisingly close to those of Copernicus's spheres. Hence, he envisioned the cosmos as designed according to a precise geometrical scheme: the sphere of Saturn circumscribed a hexahedron, which circumscribed the sphere of Jupiter, which circumscribed a tetrahedron, which circumscribed the sphere of Mars, which circumscribed a dodecahedron, which circumscribed the sphere of Earth, which circumscribed an octahedron, which circumscribed the sphere of Venus, which circumscribed an icosahedron, which finally circumscribed the sphere of Mercury (fig. 7.6).

Kepler did not suppose that there were giant material polyhedra looming in the sky. Rather, he conceived of the solids as Platonic forms, entities imperceptible to the senses but intelligible to the mind, which determined the number and relative sizes of the spheres by means of their geometrical properties. The actual celestial spheres were, however, material entities—at any rate, material enough to physically carry the celestial bodies in their revolutions.

Fig. 7.6. Kepler's theory of Platonic solids and planetary spheres, as illustrated in *Mysterium Cosmographicum*, 1596.

Just exactly how material the celestial spheres could be had become a matter of debate. The revival of Ptolemaic astronomy meant also a renewal of the practice of astronomical observation, and hence a growth of empirical astronomy in the second half of the sixteenth century. While new data relative to the cyclical motions of planets, Moon, and Sun were being recorded, a series of unexpected astronomical events began casting doubts on the reality of the spheres. From November 1572 to March 1573, the Danish astronomer Tycho Brahe observed a bright object appear in the constellation of Cassiopeia. The object shone for a few months and then disappeared. Brahe concluded from his observations that this new phenomenon was not a sublunar object but indeed a *stella nova* ("new star"), in open contradiction of the old belief that the skies were immutable.[8] Furthermore, in 1577 he observed and carefully traced the trajectory of a comet. His measurements revealed that the comet could not be an atmospheric effect, as the Aristotelians believed, but was a superlunar body traveling on a highly oblique and eccentric orbit that traversed the spheres of Venus and the Sun. Other new stars and comets were also noticed in the following decades, and the observations confirmed ever more confidently that they were not isolated supernatural portents, marvels carrying special messages from God, but rather regular natural occurrences amenable to the same

[8]Modern astronomers have determined that Brahe's "nova" was an exploding star, a rare phenomenon (rarer still if it has to be visible to the naked eye) that is today called a supernova. The 1572-1573 supernova has been identified with a type I supernova (maximum brightness), and its residual radiation is still detectable as infrared and radio waves.

kind of scientific explanation as to the other astronomical phenomena. Accommodating them within astronomical theory, however, required at least some radical changes to the old-style, snugly fitted cosmos. Brahe was not a revolutionary thinker; he could not even accept Copernicus's moving Earth. Accustomed to conceiving of the cosmos as filled everywhere with matter of one kind or another, he regarded the immense distance that Copernicus had been forced to suppose between the Earth and the stars as preposterous. According to traditional mentality, it would have been senseless wastefulness on the part of God to create a universe disproportionally larger than Earth, the largest part of which would be without any apparent purpose. Nevertheless, Brahe's observations convinced him that traditional astronomy was in need of a radical reform. He formulated his own alternative model, aiming at a *via media* ("middle way"), an optimal compromise between the reassurance of geocentrism and the modernizing flair of heliocentrism. In his model, the Earth was stationary at the centre (or in the proximity thereof) of the sphere of the stars, and the Moon and the Sun rotated around it, while the planets rotated around the Sun (fig. 7.7).

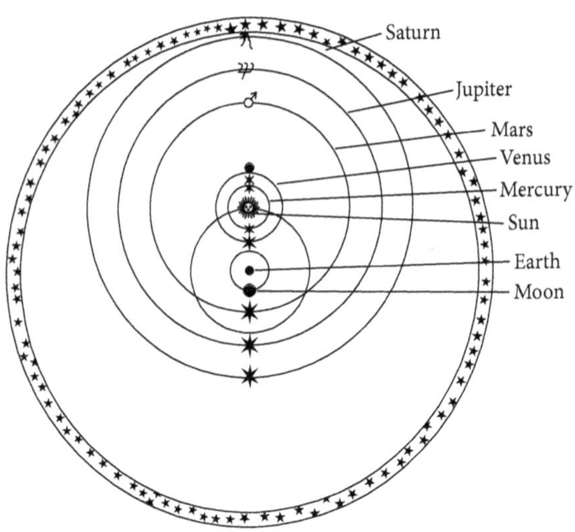

Fig. 7.7. Tycho Brahe's geo-heliocentric model, also known as "Tychonic system" or "Tychonian system". (Source: Tycho Brahe, *Introduction to the New Astronomy*, 1588).

Brahe preserved the principle of circularity and uniformity but discarded the solid spheres. Since in his model the path of the Sun intersected that of Mars, the spheres could no longer be impenetrable physical objects. Like other astronomers of his time, he was open to the possibility that celestial matter was a fluid through which the planets moved like fish in water. The road was open for the next generation to reduce the role of the spheres to that of immaterial forms, or even jettison them altogether. If this outlook was combined with a belief in heliocentrism, where the diurnal rotation was attributed to Earth, it became possible to imagine that the sphere of the stars was infinitely large (fig. 7.8). In the modernizing cultural currents of the time, some radical thinkers, as for instance the philosopher Giordano Bruno, began associating the new astronomy with the daring vision of an open, centre-less universe populated by infinite heliocentric systems.

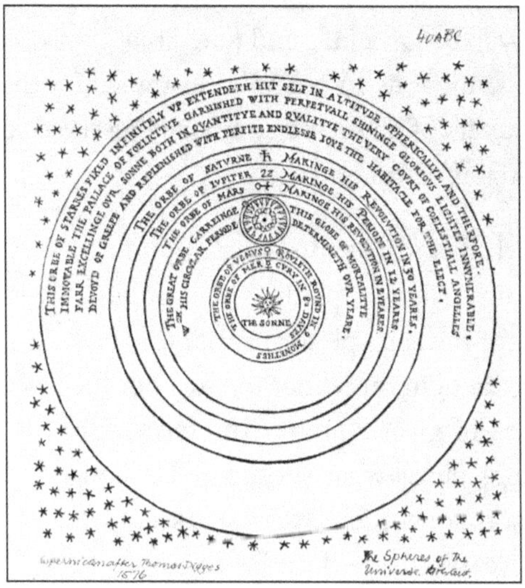

Fig. 7.8. The heliocentric system in an infinite universe.

Brahe was not the only astronomer seeking a *via media* between Ptolemy and Copernicus. In fact, studying the mathematics of *De revolutionibus*, astronomers realized that, as far as mathematical features were concerned, the difference between a geocentric and a heliocentric amounted to no more than a change in point of view. Not only were the two models empirically equivalent, they could also be transformed into one another by geometrical operations. Playing with points, lines, and circles on paper, there is no substantial difference between making point A revolve around point B, and making point B revolve around point A. Astronomers began exploring variations that mixed the features of the two models. For instance, a model in which the inferior planets orbit the Sun, while the Sun and the superior planets orbit the Earth was championed by a rival of Brahe's. Before a consensus could be reached on which variant should be chosen over the others, astronomers had to come to an agreement on the criteria for the choice. Was there a model superior to any other in empirical accuracy, or was it always possible to have multiple models that saved the phenomena equally well? And if empirical accuracy offered no basis for choice, what principles were to be followed, and who decided them? Brahe recruited the young upstart, Kepler as his assistant in the controversy, even though he knew Kepler to be a Copernican. Charged with the task of writing a defence against Brahe's rival, Kepler astutely bypassed the question of which mathematical model was the correct one, and focused instead on the need of physical premises to distinguish between mathematical models. He rebuked his rival for being a mathematical instrumentalist and for mistakenly believing Copernicus to have held the same position. In contrast, he revealed that the foreword to *De revolutionibus* had been written by Osiander, and noted that the book actually expressed realist views. Upholding Copernicus as an example, he argued for the right of an astronomer to be concerned with matters of physics, traditionally reserved to philosophers—to be, in his words, a "philosophizing astronomer":

> The astronomer ought not to be excluded from the community of philosophers who inquire into the nature of things. He who predicts as accurately as possible the movements and positions of the stars performs well the duty of the astronomer. But he who employs true opinions about the form of the universe does it better, and is held worthy of greater praise. The former, to be sure, draws conclusions that are true as far as what is observed is concerned; the latter not only does justice in his conclusions to what is seen, but also embraces the innermost form of nature in order to draw his conclusions. (Kepler, *Apologia Tychonis contra Ursum*, 1600)

Kepler argued that an astronomer should not be content with just any hypothesis that saves the phenomena, but ought to look for the one that represented the "innermost form of nature". In addition to this, he believed that a seemingly accurate hypothesis could be proven false by showing that it was not accurate in the details. If as an astronomer he felt entitled to philosophize, as a mathematician he was still bound to submit his hypotheses to the test of observation in the most stringent way possible. It is worth noting that on this point Kepler was departing from Plato, according to whom a true astronomer should not limit himself to with "the starry heaven which we behold", but should seek the "true realities" of the heavenly motions in "pure numbers and perfect figures."[9]

[9]See Sect. 6.1.

When Brahe died suddenly in 1601, Kepler replaced him as the Imperial Mathematician at the court of Rudolph II in Prague, and gained access (not before some wrestling with Brahe's family) to the body of data that Brahe had collected. Thus, he set out to fine-tune Copernicus's planetary models to fit the data "as accurately as possible", initially hoping to fix the discrepancies between the observations and the Platonic architecture that he believed to constitute "the innermost form of nature". He concentrated on the orbit of Mars, for which the largest number of observations were available. But after eight years of calculating and comparing, in a struggle that he described as "the war against Mars", he had to accept a disappointing conclusion: neither Copernicus's combination of eccentric and minor epicycle, nor a resurrected Ptolemaic equant could reproduce the positions of the planet within less than eight minutes. Since Ptolemy's time, astronomers had settled for a theory-observation agreement of ten minutes at best. But, Brahe's data set a new standard, as they were more precise than eight minutes. Kepler was determined that such unprecedented precision should be honored by an adequately matching model. In the end, his confidence in Brahe's observational ability led him to reject not only the celestial spheres and hence his polyhedral architecture, but ultimately also the venerable principle of uniformity and circularity. As he put it, "These eight minutes alone led the way to the reformulation of all of astronomy." (Kepler, *Astronomia Nova*, 1609.)

Kepler narrated his quest and his results in his aptly titled book *Astronomia Nova* ("New Astronomy"). Having abandoned the physical assumption that had been until then the guiding principle of astronomy, he replaced it with his own physical ideas. He no longer believed in the existence of spheres of crystalline celestial matter, eternally rotating according to their own nature, and carrying the planets in their motions. He assumed instead that the planets would tend to stand still by their nature, and that some power or force was needed to move them along their paths. Inspired by Renaissance Neoplatonism, Kepler believed that this moving power emanated from the Sun. As the study of magnets was in fashion among natural philosophers in the early seventeenth century, he toyed with the idea that the Sun exercised a kind of magnetism capable of alternately attracting and repelling the planets, so as to propel them along their orbits. The reason why a planet did not move uniformly but changed speed according to the distance, faster at the perihelion (the point of minimum distance from the Sun) and slower at the aphelion (the point of maximum distance), was that it felt the Sun's power more intensely when it was closer than when it was farther. Kepler speculated that the moving power of the Sun was inversely proportional to the distance. Using this guiding assumption, he carried out painstaking calculations and concluded that the motion of Mars was such that the radius from the Sun to the planet swept equal areas in equal times (fig. 7.9).[10]

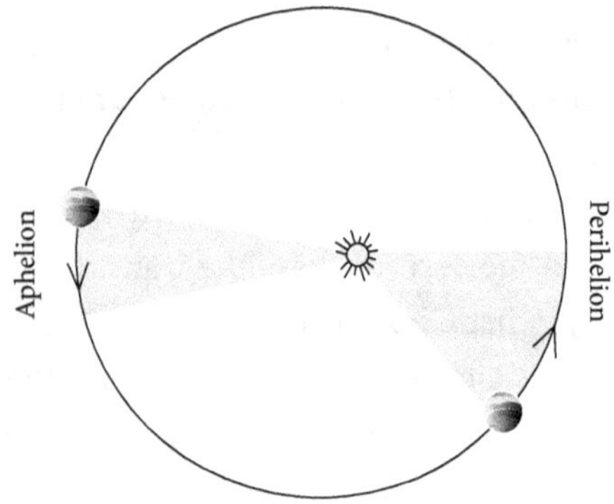

Fig. 7.9. Kepler's mathematical description of the phenomenon of inequality in the motion of a planet: the radius from the Sun to the planet sweeps equal areas in equal times. Today it is known as "Law of areas" or "Kepler's second law".

Further, after a series of trials and errors that would have discouraged a lesser spirit, he realized that the orbit was not an eccentric circle but an ellipse, with the Sun placed in one of the foci (fig. 7.10). Ellipses were known to geometers from the works of Archimedes and Apollonius as one type of "conic sections" or "conics", the family of curves that

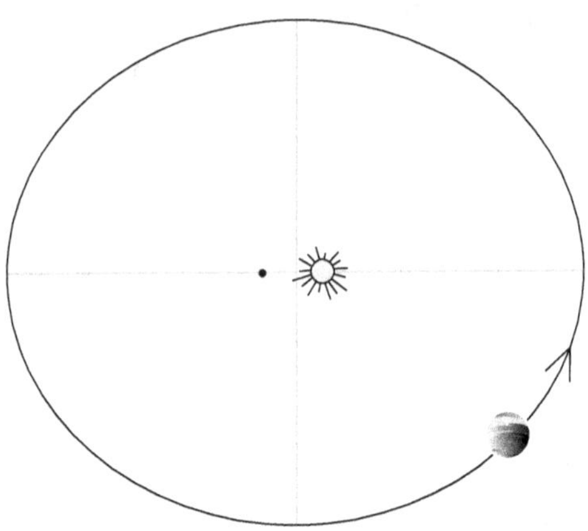

Fig. 7.10. Kepler's mathematical description of the path of planets: a planet describes an elliptical orbit, with the Sun placed in one of the foci of the ellipse. Today it is known as "Law of elliptical motion" or "Kepler's first law".

[10]Kepler's assumption that the force was inversely proportional to the distance was incorrect, but he made a compensating mistake in the computation of areas and ended up producing a correct result. Neither the correct mathematical form of the force between two masses (or, for that matter, the definition of a force) nor the correct technique for calculating areas of curved figures were known before Newton. (See Sect. 8.2 for a discussion of Newton's work.) So it came to be that Kepler arrived at the law of areas (an empirical or phenomenological law, that is, a mathematical description of observations) on the basis of an incorrect theory, and Newton at the correct theory on the basis of the law of areas. Here is a vivid reminder that in the practice of science the relationship between theory and observation is far from simple.

are obtained by sectioning a cone with a plane, but this was the first time that a use had been found for them in the natural world. Kepler had finally found a theoretical geometric entity in nature, albeit an unexpected one. The positions of Mars calculated by means of these two new assumptions—the orbit is an ellipse ("law of elliptical motion") and the radius describes areas proportional to the times ("law of areas")—displayed the desired agreement with observations, and the same was also true for the other planets.

Kepler continued his search for the mathematical forms underlying the structure of the cosmos. In a later book titled *Harmonices mundi* (1619), he turned his attention to the relation between the sizes of the elliptical orbits and the periods of the different planets. Chasing in his own creative way the Pythagorean dream of the celestial harmonies, according to which the celestial bodies emit notes according to the ratios of their orbits, he hit upon a proportion that related the planets to one another and unified them in a "harmonic" whole more literally than Copernicus's system did: the squares of the periods of the planets are to one another as the cubes of their mean distances from the Sun (i.e., the major semi-axes of their orbits.) This is known as the "harmonic law" (fig. 7.11).[11]

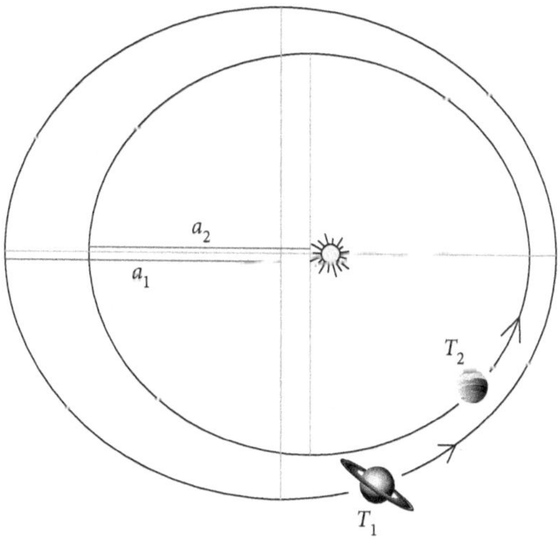

Fig. 7.11. Kepler's relation between the sizes of the orbits and the periods of the planets: the squares of the periods are to one another as the cubes of the major semi-axes, $T_1^2 : T_2^2 = a_1^3 : a_2^3$. Today it is known as "Harmonic law" or "Kepler's third law".

[11] In modern terms, Kepler's third law is, the squares of the periods are proportional to the cubes of the major semi-axes, or symbolically, $T^2 = ka^3$. Before the emergence of symbolic algebra, mathematical relationships between physical or geometrical quantities (magnitudes) were expressed in terms of proportions or equalities of ratios, and ratios were defined only between quantities of the same kind. See Sect. 4.4 for the Pythagorean theory of harmonic ratios, and Sect. 2.2 for the emergence of symbolic algebra.

Kepler's victory in the "war against Mars" was multifold: he erased the boundary between mathematics and physics, discarded the separation between terrestrial and celestial physics, and broke the spell of uniform circular motion. Although the specifics of Kepler's physics and his Pythagorean fantasy did not survive further scrutiny, his hybridization of physics and mathematics, and his mathematical descriptions of planetary phenomena passed the test of time with flying colours. Throughout the seventeenth century they spread across Europe in the works of a growing number "physico-mathematicians", a new brand of scientists who advocated a renewal of science and aimed at extending the methodology of the philosophying astronomer to all branches of natural philosophy. Kepler's results eventually reached a young physico-mathematician in Cambridge, England, by the name of Isaac Newton. Newton reformulated Kepler's descriptions of planetary motion as three "empirical laws" (numbering the first and the second in the reverse order to that in which Kepler had arrived at them), and built upon them his theory of universal gravitation. They are today known as "Kepler's laws".

Chapter 8

Mathematics and the Origins of Modern Science Part 2

8.1 The Trial of Galileo, or "How to Make the Earth Move without a Thousand Inconveniences"

An early champion of physico-mathematics was Galileo Galilei, a mathematics professor in Padua, in the Republic of Venice. Galileo became a Copernican, albeit a cautious one, early in his career, but his main interest was in the motion of terrestrial bodies. He was part of a group of Italian mathematicians inspired by the works of Archimedes, who exemplified to them an ideal combination of the rigour of a Greek theoretical geometer with the practical ingenuity of an engineer. They aspired to use mathematics in the design of buildings, instruments, and engines. Their goal was to create a theoretical science of mechanics on the model of Archimedes's studies of the equilibrium of bodies (the law of the lever and the principle of buoyancy). Galileo was also very critical of Aristotelian physics and dissatisfied with the hierarchical structure of universities. Even before taking up his post in Padua, he conducted studies of falling objects in his hometown of Pisa, in Tuscany. Aristotle had stated that the speed of a body is proportional to its weight. Galileo found that Aristotle's statement was contradicted both by logic and by observations. Making careful observations of real bodies under specially arranged and controlled conditions, and combining them with deductive reasoning—recognizably, the blueprint of modern laboratory science—to find out *exactly how* real bodies fall would become a life-long commitment for him, and would lead to his highest achievements. He found himself repeatedly embroiled in quarrels with Scholastic philosophers, and his sharp wit hardly helped his academic career. In the fall of 1609, having heard reports of a newly invented optical device capable of making things afar appear as if they were nearby, he built one such "looking glass", or spyglass, by fitting two lenses at the ends of a brass tube. To design technological novelties was one of the practical tasks for which mathematicians were valued in early modern Europe, where technology was becoming the main driving force of military power and economic progress. For mathematics professors, it was a way to gain a reputation and to supplement their meagre salaries, and Galileo had already distinguished himself as a skilled maker of mathematical instruments. Understanding the value of the spyglass for a seafaring republic like Venice, he presented his spyglass to the Venetian authorities, who rewarded him with lifelong tenure and a salary increase. His first looking glass was similar to those that other instrument makers were already putting on the market. Galileo, however, learned how to make lenses of better quality and longer focal length, and introduced other technical improvements, so that within a year he was able to produce spyglasses with 20x and 30x magnification. But his major breakthrough was to use one of his instruments to observe the sky. Until then, astronomical observations had been conducted with the naked eye. Astronomers were familiar with technical aids for aligning the eye, measuring angles, and mapping the heavenly bodies, such as rulers, astrolabes, and armillary spheres. In the sixteenth and seventeenth centuries, the production and use of these tools, driven by the needs of navigation and the popularity of the mathematical sciences among the wealthy, had reached considerable degrees of sophistication. The spyglass, however, was not simply another tool for doing old measurements better; it was a veritable game-changer. No other event in the history of astronomy had an impact comparable to that of Galileo's transformation of the spyglass into an astronomical instrument.

Galileo observed the surface of the Moon and saw that it was an earth-like body covered in craters and mountains and not at all, as traditional science and poetry would have it, a polished mirror occasionally clouded by perturbances in the Earth's atmosphere. He saw many previously unseen stars amidst the old constellations, and saw that the so called *nebulae* and the Milky Way were composed of innumerable stars that the naked eye could not resolve. He saw four unknown "planets" circling Jupiter in the same way as the Moon circled Earth. He published these discoveries in a book called *Nuncius sidereus* (*The Starry Messenger*, 1610,) which he dedicated to the Grand Duke of Tuscany. He also named Jupiters's moons "Medicean Stars" in honor of the Medici family, to which the Grand Duke belonged. This was a career move skillfully attuned to the conventions of court patronage of the time. It succeeded in its intent and gained him the favour of the Medicis. As the book quickly became a sensation among scholars and the cultivated public, Galileo was called to Florence and appointed "Philosopher and Mathematician to the Grand Duke"—a newly minted title that fit his ambitions and freed him from the academic ranks. In this way, he fashioned himself into a high-profile role model for those aspiring to redefine the boundaries of the scientific disciplines.

In *The Starry Messenger*, Galileo announced his discoveries without advancing any cosmological argument. He was aware that all the new phenomena he had discovered so far could be accommodated within a geocentric system as well as in a heliocentric one. But he became more vocal about heliocentrism shortly thereafter, when he discovered that Venus displayed a complete set of phases like the Moon, an effect that could only be explained if Venus circled the Sun. He thus began a campaign in favour of the Copernican system. Although his theoretical claims were met with predictable resistance, the experts of the Collegio Romano, the main Jesuit college, verified and approved his empirical results. This was the highest endorsement of the spyglass he could hope for. During a banquet in his honor in Rome, a classicist poet coined a dignified name for the device, "telescope", from the ancient Greek words *tele*, "far", and *skopein*, "to look". Christopher Clavius, who had been the head of the Gregorian calendar commission and was the top mathematician in Rome, rejected Galileo's Copernican argument, yet admitted that the telescopic discoveries made the old system no longer tenable. He and his colleagues at the Collegio Romano leaned toward Tycho's Brahe's *via media*, which could explain the phases of Venus without giving up the stationary Earth.

Now a celebrity, Galileo became the target of criticism by Aristotelian philosophers and by members of the clergy, who claimed that the idea of a moving Earth contradicted several passages of the Bible. He responded in two private but widely circulated letters (a common form of scholarly communication before the birth of academic journals), arguing that the Bible was not to be understood literally but needed interpretation. Theologians were competent to provide the correct interpretation in matters of faith and morality; in matters of cosmology, the expertise belonged to mathematicians. Both the Holy Scripture and the creation were expressions of God; they both needed interpreters, but they had very different characters. The Scripture was written "to accommodate the capacities of the very unrefined and undisciplined masses," and for this reason the literal meaning of its words could be different from the true meaning. Nature, on the contrary, was indifferent to human understanding and never violated the "the laws imposed upon her" (Galileo, *Letter to the Graduchess Christina*, 1615, in Galilei 2008, 115–116.) Hence, finding out about nature required a set of resources very different from theological interpretations, namely, empirical observations and deductive reasoning:

> Therefore, I think that in disputes about natural phenomena one must begin not with the authority of scriptural passages, but with sense experiences and necessary demonstrations. (Galileo, *Letter to the Grand Duchess Christina*, 1615, in Galilei 2008, 116.)

Just as Galileo promoted the fusion of mathematics and natural science, he advocated the separation of natural science from theology. In his view, there could be no conflict between science and the Bible, for the Bible was not meant to be a textbook of natural science. Quoting a cardinal, Galileo quipped, "the intention of the Holy Spirit is to teach us how one goes to heaven and not how heaven goes." (Galileo, *Letter to the Grand Duchess Christina*, 1615, in Galilei 2008, 119.)

Galileo was a polemicist too brilliant for his own good. For all the modernizing trends in science, the political climate was no longer favourable to new ideas. The Catholic Church had responded to the Protestant Reformation with a Counter Reformation, a series of reforms that aimed at regaining for the Vatican absolute control in matters of doctrine. One of these reforms was the creation of the Holy Office, also known as the Inquisition, as a doctrinal tribunal. In 1600, the ultra-Copernican Giordano Bruno had been condemned for heresy by the Inquisition and burned at the stake in a popular square in Rome. Galileo's scientific ideas were not as outlandish as Bruno's. Even so, his campaign to claim authority for mathematicians in natural science hit a sensitive spot. As his arguments started to win supporters even inside the Church, he was reported to the Inquisition. After hearings with some of the tribunal's officials, he was given a private warning *not to hold or defend as a truth* the Copernican hypothesis. The Copernican theory was declared to be a hypothesis false and contrary to the Church's doctrine, and Copernicus's *De revolutionibus*—after seventy years of being serviceable to the Church—was placed on the Index of Forbidden Books, where it remained for a few years until it was purged of any assertion of the motion of the Earth as a realistic hypothesis.

Yet, Galileo was hardly deterred. Although he refrained from defending heliocentrism explicitly, he continued to write on natural philosophy and its methods, now switching to the Italian language to reach a wider audience. He was bolstered by the *Accademia dei Lincei* ("Academy of the Lynx-Eyed"), an association of aristocrats who supported his pioneering of experimentation and provided the first example of a new kind of venue for scientific research, the scientific academy. Continuing on the mission to fuse together mathematics and natural philosophy, Galileo penned the now famous metaphor of nature as a book written in mathematical language,

> Philosophy is written in this all-encompassing book that is constantly opened before our eyes, that is the universe; but it cannot be understood unless one first learns to understand the language and knows the characters in which it is written. It is written in the mathematical language, and its characters are triangles, circles, and other geometrical figures; without these, it is humanly impossible to understand a word of it, and one wanders around pointlessly in a dark labyrinth. (Galileo, *The Assayer*, 1623, in Galilei 2008, 183.)

When a fellow member of the *Accademia dei Lincei* was elected pope with the name of Urban VIII, Galileo loosened his restraint and published a book titled *Dialogue On the Two Chief World Systems, the Ptolemaic and the Copernican* (1632), hoping naively that the new pope would appreciate it. The book presented three characters debating the values of the geocentric and heliocentric theories, Salviati, Galileo's persona, Simplicio, an Aristotelian, and Sagredo, an intelligent and supposedly neutral participant.[1] Formally, the *Dialogue* respected the injunction to present both theories as hypotheses. In the substance, however, it was a brilliant defence of the Copernican ideas and a demolition of the Aristotelian-Ptolemaic views. Within a year, Galileo was summoned to Rome by the Inquisition. There began one the most infamous trials in history. His enemies proved

[1] Salviati and Sagredo were the names of two friends of Galileo; Simplicius was the name of a sixth-century philosopher, but it was also a pun on the Italian adjective *semplice*, which means simple or simple-minded. The not-so-subtle double-entendre was not lost on Galileo's opponents.

to be powerful within the Vatican. A document was planted in his file, alleging that the 1616 injunction was not to discuss the motion of the Earth "in any way whatsoever". He was interrogated twice, the second time under verbal threat of torture, and was finally found guilty of "vehement suspicion of heresy"—luckily, a less harsh verdict than the full charge of heresy. He was forced to a humiliating public recantation, in which he had to kneel down and declare that he "abjure[d], curse[d], and detest[ed]" the "doctrine (which is false and contrary to the Holy and Divine Scriptures) that the sun is the center of the world, and that it does not move from east to west, and that the earth does move and is not the center of the world[.]" ("The Crime of Galileo. Indictment and Abjuration of 1633", in *Internet Modern History Sourcebook*, Paul Halsall, ed., at http://www.fordham.edu/Halsall/mod/modsbook.asp.)[2] He was sentenced to lifetime imprisonment, but the sentence was soon commuted to house arrest. He spent the rest of his days in poor health in his villa near Florence. Yet, he did not stop working, and wrote his most influential book, *Discourses and Mathematical Demonstrations on Two New Sciences*—another dialogue between Salviati, Simplicio, and Sagredo—which his friends smuggled across the border and published in the Netherlands in 1638.

In the *Dialogue*, Galileo broached the problem that Copernicus had left open. How can it be possible for the Earth to move, when we do not observe any effect of that motion? If the Earth were twirling around its axis at the speed required by the daily rotation, any objects not safely attached to the ground would be ejected away from the center.[3] Birds and clouds would be left behind, and there would be a constant westward wind. A projectile thrown upward would not fall back to the same spot because the ground would be shifting eastward during the fall—or so the objections went since Aristotle's time. In Simplicius's sarcastic words, "The important point is to be able to make the Earth move without a thousand inconveniences." (Galileo, *Dialogue on the Two Chief World Systems*, 1632, in Galilei 2008, 211.) For two thousand years this commonsensical reasoning had been almost universally regarded as conclusive proof that the Earth was stationary. Galileo articulated a powerful counterargument. Using the analogy between the Earth and a ship, he argued that a person standing on the deck of a ship could draw

[2] According to an unverifiable legend, as Galileo stood up after the abjuration, he muttered, "*Eppur si muove!*" ("And yet it moves!".) This phrase became popular as an expression of defiance against powerful but untruthful dogmas.

[3] Galileo focused on the daily rotation, even though his counterargument applies to the annual revolution as well. More specifically, he focused on what he saw as the strongest objection, namely, the belief that a dropped object would be seen to fall behind.

no conclusion concerning the motion of the ship simply from observing the fall of stones dropped from the mast. For—contrary to what people believed on prejudice, without bothering to verify it—the stones would always fall along the mast and end at the foot of it, regardless of whether the ship was at rest or moving. This happened for two reasons. First, once horizontal motion was impressed upon a body, as would be the case for any object carried on a ship, the body would have a natural tendency to remain in motion in the same direction and with the same speed unless some impediment occurred to retard or change it. Second, the natural tendency of the body to fall under the effect of gravity would not constitute an impediment, since it would act in the perpendicular direction. Both reasons contradicted Aristotelian physics, according to which a heavy body like a stone would require to be forcefully pushed or pulled to move in any direction other than toward the centre of the Earth, and the natural tendency of the body to fall under the effect of gravity would overtake and destroy any other motion. In open contrast with Aristotle, Galileo argued that the overall motion of a terrestrial body can be the result of two motions that do not interfere with one another, a vertical motion to or from the centre, and a horizontal motion along a surface that does not slope either upward or downward. The Aristotelian view that horizontal motion requires a moving force derived from everyday experience with real bodies. But Galileo explained that the slowing down and disappearing of motion was not spontaneous. It was an effect of "accidental impediments", such as friction and the resistance of air, or the ball hitting something, breaking down, and other real-life occurrences. As a demonstration, he invited the reader (in the book, Salviati invited Simplicius and Sagredo) to perform a thought experiment, namely, to imagine letting a perfectly round and smooth bronze ball roll down on an inclined plane "polished like a mirror" from a given height, and then up a second, equally "fastidiously polished" plane (Galileo, *Dialogue on the Two Chief World Systems*, 1632, in Galilei 2008, 229-230) (fig. 8.1).[4]

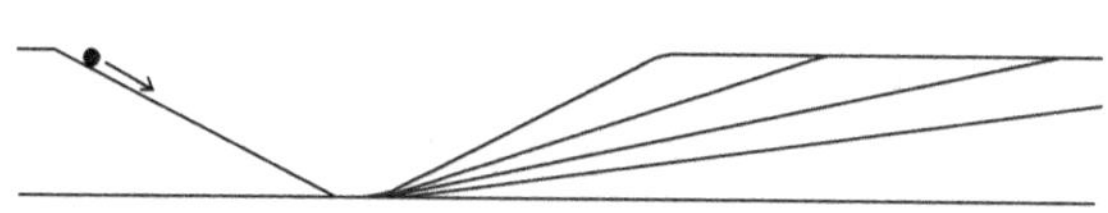

Fig. 8.1. Galileo's thought experiment of an ideal ball rolling on two ideal inclined planes.

[4] "Thought experiment" is a modern term for this kind of hypothetical reasoning on idealized empirical cases. Note that a thought experiment is not one of the two resources that Galileo advocated, deductive reasoning and experiment, but is a hybrid the two.

Rolling downward, the ball would accelerate, until it would reach the bottom with a certain speed. Rolling upward it would decelerate, until it would come to rest at the same height as that from which it started. The duration of the motion and the distance traversed would depend on the inclination of the second plane. Specifically, they would increase as the inclination decreased, so that in the extreme case when the second plane became parallel to the horizon, the ball would neither accelerate nor decelerate, but would continue rolling indefinitely with constant speed. On a frictionless horizontal surface there would be no cause for retardation or acceleration. Just as a body initially at rest would stay at rest, a body initially in motion would continue moving without change. "If such surface were endless, the motion on it would likewise be endless[.]"Similarly, a rock falling from the mast of a moving ship would continue to move on a horizontal path with a motion "indelibly impressed upon it" (Galileo, *Dialogue on the Two Chief World Systems*, 1632, in Galilei 2008, 231.). The rock would follow the ship even as it acquired vertical motion due to gravity, and it would end at foot of the mast. Hence, an observer on the ship would not be able to decide whether the ship were at rest or sailing smoothly just by observing the falling stone. If this was the case for an unnatural motion such as that of a sailing ship, with greater reason it would be the case for a rotating Earth, to which the diurnal rotation would "belong naturally" (Galileo, *Dialogue on the Two Chief World Systems*, 1632, in Galilei 2008, 226.) Furthermore, the same rotation would belong naturally to every part of the terrestrial sphere, air included, and consequently light objects would not be impeded by the resistance of air but rather carried along with it. Therefore, no terrestrial observer would be able to conclude that the Earth is still just by the observation that falling things are not being left behind.

Galileo turned the age-old objection against the motion of the Earth on itself by means of two new principles, the principle of conservation of motion and the principle of composition of motion. Aristotelian physics regarded terrestrial motion as an occurrence of change, a process which would sooner or later come to an end according to the physical nature of the body. Even Kepler understood motion in this way. Extending terrestrial physics to the heavens, he assumed that if the power of the Sun were switched off, the planets would suddenly stop on their orbits. This understanding presupposed a categorical distinction between motion (a process, requiring a cause) and rest (the state at the end of any process, after all causes of change are removed). In contrast, Galileo's idea that motion *continues* when there is no cause for acceleration or deceleration reduced motion to a state similar to the state of rest. A body on a surface equidistant from the centre was "indifferent" to its motion, and would change its state only under the action of an external force. This idea provided the basis for understanding motion and rest as *relative states*. As Galileo further explained, no experiment conducted inside a windowless cabin inside the ship would be able to reveal whether the ship is anchored at the dock or sailing smoothly on the ocean. Every object in the ship is in motion or at rest relative to the ship, while the ship and all its cargo are in motion or at rest relative to the dock. Today, this notion of the relativity of motion is called classical or Galilean relativity. Galileo's principle of "indifferent" motion was a precursor of the principle of inertia. It referred to circular motion around the centre of the Earth, which is only approximately straight because the radius of the Earth is large on the scale of human observation. For this reason, Galileo's version of the principle is today called "horizontal inertia". It was a transitional idea, a bridge from the old era to the new era of science. On one side, it stemmed from the old understanding of uniform circular motion as natural and eternal, and from the old understanding of space as structured in distinct spheres, each having its own centre. On the other side, it led to the modern view of motion as a relative state. Soon, Descartes and others would formulate the principle of inertia for motion on a straight line, or rectilinear inertia, which implies a homogeneous space with no centre. Newton would then adopt rectilinear inertia as the first law (or axiom) of his mechanics.

For Galileo, motions did not overcome one another according to the "nature" of the moving body, but they compounded geometrically and in the same way regardless of the body's nature. Like other physico-mathematicians, he envisioned a new kind of physics, in which the phenomena would be reduced to mathematically describable attributes and laws. He argued that the sensible qualities of bodies, such as hotness, colour, taste, softness, and so on, did not constitute the nature of the body but were in fact just "secondary qualities", which did not inhere to the sensed object but to the sensing subject. They were produced in the senses by the "primary qualities", or "real attributes" of the object, and these were the mathematical attributes of "sizes, shapes, quantity, and slow or fast motions." (Galileo, *The Assayer*, in Galilei 2008, 187). This idea was further developed by Descartes and others into an intellectual project called mechanical philosophy, which aimed at understanding nature as a machine, an assemblage of particles of matter in motion according to mathematical laws. In the vision of mechanical philosophers, the science of mechanics was no longer just the study of artificial machines but became the fundamental natural science, the science of motion.

In his last book, *Discourses and Mathematical Demonstrations concerning two New Sciences*, also known as *Two New Sciences*, Galileo laid the foundations of a geometrical representation of motion that was one of the lasting results of mechanical philosophy, and constituted the precondition of Newtonian mechanics. Having discarded the Aristotelian idea that the speed of fall is proportional to the weight, volume, or density of the falling body, he noted that simple observations showed that the free motion of a heavy falling body was a continuously accelerated motion. Nobody had ascertained, however, exactly how a falling body accelerated. Thus, he set out to find the mathematical law describing free fall. He combined mathematical reasoning and detailed empirical observations, as Kepler had done. Unlike Kepler, however, Galileo had to uncover the mathematical form underlying the physical phenomenon by stripping away the "accidental impediments" that affected every real occurrence. For this reason he conducted his observations on specially prepared setups. He reduced the accidental impediments to a minimum and controlled the conditions of measurement, in such a way as to approach the ideal manifestation phenomenon as closely as possible. Marrying the rigour of the geometer with the practical skills of the engineer, he pioneered the modern practice of laboratory experimentation. This is an active kind of empirical observation, in which physical entities are manipulated with the purpose of approximating the mathematical models of the phenomena under study.[5] He conducted a series of experiments with heavy balls rolling along inclined planes, taking care to minimize friction by making the balls "hard, smooth, and very round", and letting them roll inside "very straight, smooth, and polished" wooden grooves lined with parchment "also as smooth and polished as possible", at different angles of inclination (Galileo, *Two New Sciences*, 1638, in Galilei 2008, 350.) With repeated measurements, he was able to confirm that the distances traversed were to each other as the squares of the times needed to traverse them. Through mathematical analysis, he demonstrated that this mathematical relation was a distinctive property of *uniformly accelerated motion*, that is, of a motion in which speed acquires equal increments in equal intervals of time. He represented motion geometrically by letting a line AB represent the total time of fall, and lines perpendicular to AB represent the speeds at each instant. In this way, uniformly accelerated motion would be represented by a triangle having the total time as height and as the final speed as basis. Also, he let the distances traversed be represented by segments of a line HI. He was thus able to demonstrate also

[5] An experiment is a recursive process, in that the mathematical models are in turn refined and modified to accommodate what is learned during the manipulations.

that the successive distances, HL, LM, MN, etc. traversed in equal time intervals, AD, DE, EF, etc. as the speed increased uniformly, were "to one another as the series of odd numbers" 1,3,5, etc. (Galileo, *Two New Sciences*, 1638, in Galilei 2008, 348) (fig. 8.2).

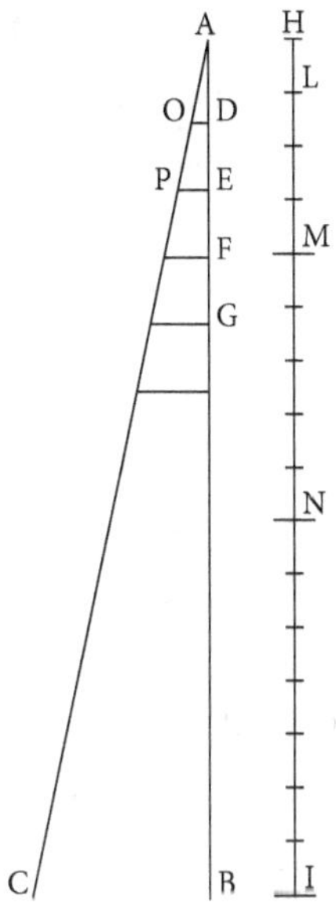

Fig. 8.2. Galileo's geometrical representation of uniformly accelerated motion in *Two New Sciences*, 1638. The line AB represents time, and the lines OD, PE, etc. represent the uniformly increasing speeds. The lines HL, LM, MN, etc. represent the distances traversed in the equal intervals of time AD, DE, EF, etc.

(Source: Galileo, *Two New Sciences*, 1638.)

Galileo went on to apply his law of falling bodies (i.e, that the distances traversed by a falling body are as the squares of the time intervals), his principle of indifferent motion, and his principle of composition of motions to deduce the trajectory of a projectile. As he wrote, "It has been observed that missiles and projectiles describe a curved path of some sort. However, no one has pointed out that this path is a parabola." (Galileo, *Two New Sciences*, in Galilei 2008, 334.) Like Kepler ellipse, the parabola was one of the conic sections studied by Apollonius as a subject of theoretical geometry, but it had hitherto never been used in natural science. Galileo explained that if a body was projected horizontally with a given speed and then was allowed to fall freely, it acquired at the same time a uniform horizontal motion and a uniformly accelerated downward motion. The overall motion would therefore be compounded of these two perpendicular motions. He then represented as lines the distance traversed by the two motions in equal intervals of time along the respective directions. Using the previously demonstrated properties of uniform motion and uniformly accelerated motion, he showed that the resulting path would be a semi-parabola having a vertical axis and the vertex at the starting point (fig. 8.3.)

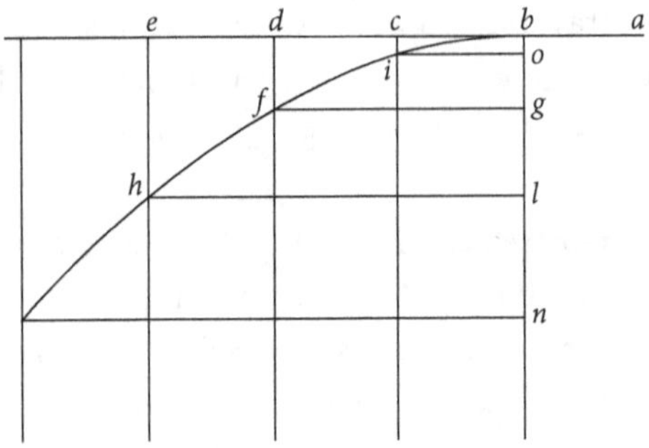

Fig. 8.3. Galileo's geometrical representation of the semi-parabolic trajectory of a projectile. (Source: Galileo, *Two New Sciences*, 1638)

Galileo expressed the mathematical relations between physical quantities verbally as proportions between magnitudes, and graphically by means of lines representing time intervals, speeds, and distances. He did use numerical measurements of time intervals and distances to verify the mathematical relations. Nonetheless, he remained within the Greek theoretical tradition and did not identify the physical magnitudes with their measured numerical values. Just as Galileo in Italy was beginning to lay the foundations of a mathematical description of motion using traditional mathematics, other mathematicians in Europe were starting a great renewal of mathematics. In the Netherlands, Simon Stevin had just extended the definition of numbers to include continuous magnitudes, and Descartes was developing a general symbolism to represent relations between numbers (in Stevin's enlarged sense) in the form of algebraic equations. Descartes developed also a method to represent all the algebraic quantities as lines. Moreover, he and, independently, another Frenchman named Pierre de Fermat showed that equations could be represented as curves, thus achieving a unification of algebra and geometry that took the name of analytic geometry.[6] As a result of these developments in arithmetic and algebra, today we are able to

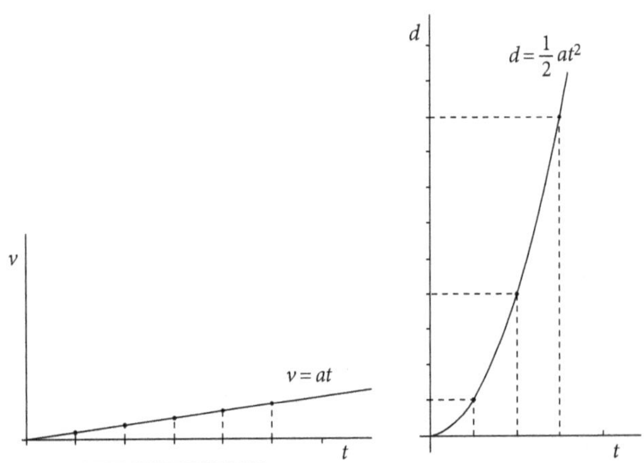

Fig. 8.4. Modern Cartesian representation of uniformly accelerated motion.

[6]See chap. 2 for a discussion of these developments.

represent uniformly accelerated motion by means of the equations $v = at$ and $d = \frac{1}{2}at^2$ (where t is the time, d the distance, v the speed, and a the acceleration), and thanks to the creation of analytic geometry we can represent the equations by means of Cartesian diagrams (fig. 8.4).

8.2 Newton and the Mathematical Principles of Natural Philosophy

Isaac Newton was a twenty-two-year old student in Cambridge when he taught himself analytic geometry reading a translation of Descartes's *La Géométrie.* He went on to study all the works he could find in the field that was then called analysis; among these were the collected works of Viète and a book by John Wallis titled *Arithmetica infinitorum* ("The Arithmetic of Infinites"). In two years, from 1664 to 1666, he mastered analysis and then proceeded to invent calculus. Analysis—widely believed to be a re-discovered method of the Greeks—had developed into the quest for a method to use algebra and analytic geometry to find tangents and areas of curves. Wallis and others (we may mention Bonaventura Cavalieri, a disciple of Galileo, and Fermat, the co-inventor of analytic geometry) were in the process of developing a method that made use of ratios of "infinitesimals", that is, "infinitely small" lines for the tangents, and of sums of infinite numbers of infinitesimal rectangles for the areas. Building upon their result, Newton pictured curves as described by moving points. He thus reduced the problem of finding the inclination of the tangent to a problem of motion, specifically, the problem of determining the rate of "flow", or instantaneous speed, of the point. He called the variable distance traversed by the moving point in a given direction the "fluent" and the rate of flow the "fluxion"; they correspond to what we call, respectively, "variable" and "derivative". He developed a general method of fluxions and areas, and demonstrated the general theorem of calculus, according to which the calculation of the fluxion of a curve is the inverse of the calculation of the area. Newton left his results unpublished, until they were independently found and then published by the German Gottfried Wilhelm Leibniz in the 1680s. There ensued one of the most bitter priority disputes in the history of science, which dragged on well into the eighteenth century, needlessly and unproductively pitching Newton's English supporters against the continental followers of Leibniz. It is a lesson in the irony of history that Newton's crowning achievement in physics, his rational mechanics, was embraced and greatly expanded by the continental mathematicians, who made fruitful use of Leibniz's version of differential and integral calculus.

Strange as it may sound today, there was no advanced mathematical teaching at Cambridge until the very end of Newton's student years. The first mathematics professorship, named Lucasian Chair of Mathematics, was instituted in 1663, and the first Lucasian professor was Isaac Barrow, one of the pioneers of the method of infinitesimals. Barrow eventually noticed the talent of the young Newton, who had become a Fellow at Trinity College after graduation. In 1669 Barrows left Cambridge and appointed Newton as his successor. Newton kept this post for almost thirty years. He continued to produce deep and original results in mathematics, optics, mechanics, and astronomy, and became the most famous mathematician of all times (making the Lucasian Professorship, to this day, the most prestigious mathematics position in the world.) Here, we will take a glance at how his theories of motion and of universal gravitation completed the Copernican Revolution and firmly established mathematics at the foundation of modern science.

During his largely independent training, Newton absorbed the results of the physico-mathematicians, in particular, of Kepler and Galileo, as well as their methodological lesson about the centrality of mathematics in natural philosophy. He deliberately titled his masterpiece, *Philosophiae naturalis principia mathematica* (in English, *The Mathematical Principles of Natural Philosophy*) and composed it on the model of a traditional geometry treatise, starting with first principles (definitions and axioms) and proving from them a series of theorems. In the introduction, he explained his goal to create a new science of nature founded on mathematical laws, thereby different from the Aristotelian science of "substantial forms" as well as from more recent theories that postulated "occult qualities". As it was customary, he called upon "the ancients" to prop his argument:

> Since the ancients (as we are told by *Pappus)* esteemed the science of mechanics of greatest importance in the investigation of natural things, and the moderns, rejecting substantial forms and occult qualities, have endeavoured to subject the phenomena of nature to the laws of mathematics, I have in this treatise cultivated mechanics as far as it relates to philosophy. (Newton "Author's Preface to the Reader", *Principia*, 1687, in Newton 1995, 224.)

Newton deferred to the authority of Greek mathematics because he was aware that mechanics did not enjoy the status of science but was rather considered to be a "manual art", therefore an imprecise and inaccurate endeavour. He argued that it needed not be so. Just as mathematicians had lifted geometry from the coarse hands of "artificiers" and elevated it to a science, they could transform practical mechanics into the science of "rational mechanics". Geometry was restricted to the magnitude of bodies; mechanics referred to their motion. Hence, rational mechanics would be "the science of motions resulting from any forces whatsoever, and of the forces required to produce any motions, accurately proposed and demonstrated." (Newton, "Author's Preface to the Reader", *Principia*, 1687, in Newton 1995, 224.)

The concept of force, which had been loosely and unsystematically employed by mechanical philosophers, became central to Newton's mechanics. Newton struggled to define it precisely and to distinguish it from other physical quantities as quantity of motion and inertia. Echoes of the struggle are found in the definitions that he finally published in the *Principia*. There, initially he called inertia the "innate force of matter," and defined it as "a power of resisting", only to go on and specify that it "differs nothing from the inactivity of the mass" and "may, by a most significant name, be called inertia." He then distinguished inertia from "impressed force", the latter being "an action exerted upon a body in order to change its state, either of rest, or of moving uniformly forward in a right line." Today, physicists reserve the term force to motion-changing action. They also accept that no single definition could suffice to fix the meaning of the term, just as no single definition could completely define mass (Newton tried with "mass is the quantity of matter, defined by the bulk and density conjointly"). To define these quantities it is necessary to refer to the mathematical relation between force, mass, and change of motion in the whole (Newton, *Principia*, 1687, in Newton 1995, 227.) This relation was expressed by Newton in the form of three "Axioms, or Laws of Motion" (Newton, *Principia*, 1687, in Newton 1995, 233-234):[7]

> *Law I.* Every body continues in its state of rest, or of uniform motion in a straight line, unless it is compelled to change that state by forces impressed upon it.
>
> *Law II.* The change of motion is proportional to the motive force impressed, and is made in the direction of the right line in which that force is impressed.

[7] Although the mathematical relation between force, mass, and acceleration continues to be commonly articulated in three distinct laws, the second law alone suffices to express it completely, since the other two laws can be derived from it. The second law was first expressed as the equation now taught in every physics course, $F=ma$, by a follower of Leibniz, the Swiss mathematician Jacob Hermann, in 1716.

> *Law III.* To every action there is always opposed an equal reaction; or, the mutual actions of two bodies upon each other are always equal, and directed to contrary parts.

While the principle of rectilinear inertia had already been sketched by mechanical philosophers, the force-mass relation was Newton's key achievement.

Newton included in the definitions also a special kind of force, which he knew to be essential to the explanation of planetary motions, that is, a force that pulls bodies toward a fixed point and that he called "centripetal force". Kepler had maintained that planets were kept in motion by a "power" emanating from the Sun. In his conception, this solar power was needed to push the planets along their orbit, for otherwise the planets would stop and stand still. Galileo had limited his studies of motion to horizontal inertia and the fall of heavy bodies, both of which he regarded as "natural" and hence not in urgent need of a more specific cause.[8] Other mechanical philosophers explored the confusing idea that a body revolving around a centre is subjected to a "centrifugal force", that is, a force "fleeing the centre." Newton realized that no force is needed to keep a planet moving, but a force is needed to curve its path, and it has to be directed toward the center of curvature. Similarly, he understood that the acceleration of a falling body implies a force directed toward the centre of the Earth. He thus coined for such forces the term "centripetal", which means "seeking the centre".

Newton derived a number of theorems from the three laws, demonstrating the properties of motion in general, and of motion under the action of a centripetal force in particular. The theorems were clearly intended to reproduce Galileo's laws of fall as well as Kepler's laws of the planets.[9] Famously, in his later years Newton recounted the moment in which he had the intuition that both sets of laws were reducible to the same cause. He told his first biographer that he was in his mother's garden during a break from university (he had had to leave Cambridge because of the plague) when the "notion of gravity came to his mind [...] occasion'd by the fall of an apple, as he sat in a contemplative mood" (W. Stukeley, *Memoirs*

[8] Galileo did point out that there existed various opinions among philosophers "as to the cause of the acceleration of natural motion". But he stated that, for the time being, his purpose was "merely to investigate and demonstrate some of the properties of accelerated motion (whatever the cause of this acceleration may be)". (Galileo, *Two New Sciences*, in Galileo 2008, 340.) After Newton's clarification of the role of force, Galileo's results was classified as kinematics, the part of mechanics that is concerned only with the description of motion, and Newton's results as dynamics, the part that involves the causes of motion. It is worth to point out, in this regard, that the emergence of the new science in the seventeenth century implied a deep revision of the question of what is a "cause" of natural phenomena, and hence of what constitutes an explanation in natural science.

[9] The term "law" is used in science with different meanings, and the meanings can change over time.

of Sir Isaac Newton's Life, 1752, at *The Newton Project*, R. Iliffe and S. Mandelbrote, eds., http://www.newtonproject.sussex.ac.uk.) Supposedly, the sight of the falling apple sparked his realization that the force of gravity ("heaviness") that attracts heavy objects toward the centre of the Earth could be the same force that makes the Moon revolve around the same centre. Although the apple story is probably a retrospective fabrication, it aptly illustrates one dimension of the "Newtonian synthesis", that is, the unification of terrestrial physics and astronomy under the same mechanical principles. Newton explained his idea in a non-mathematical summary of the *Principia*, which was published after his death with the title, *Treatise of the System of the World*. If a stone is thrown horizontally from the top of a mountain, it will fall to the ground describing a curved trajectory as a result of the composition of motions, as clarified by Galileo.[10] The greater the speed of the throw, the farther the stone will go before it hits the ground. When the initial speed is sufficiently high, the stone will complete a full circle around the Earth and return to its starting point. Neglecting the resistance of air, the horizontal component of the motion is unaffected by any force; therefore, it will continue unchanged, and the stone will go on circling the Earth in the same way as the Moon does, and in the same way as the planets circle the Sun and Jupiter's moons circle Jupiter (fig. 8.5).

Fig. 8.5. Illustration of Newton's intuition of universal gravitation, freely inspired by the picture in Newton's *The System of the World* (1728)

[10]Galileo analyzed the motion of projectiles as composed of a uniform horizontal component, which he believed to be circular but approximated as a rectilinear motion, and a uniformly accelerated vertical component. Newton assumed instead a uniform inertial component, rectilinear and tangential, that is, perpendicular to the radius of the Earth, and an accelerated radial component (along the radius). The radial acceleration depended on the distance from the centre of the Earth, and it could be approximated as a constant only if the radial distance between the starting point and the end point of the motion was negligible relatively to the Earth's radius.

Newton concluded that all the phenomena in the solar system were produced by a force that he called "universal gravitation", a mutual attraction acting between any two particles of matter. In the last part of the *Principia*, also titled *The System of the World*, he specified the mathematical form of gravitational attraction: it was proportional to the masses of the particles and inversely proportional to the square of their distance. He also proved that for large spherical (or nearly spherical) bodies the force acted as if all the mass were concentrated in the center of the body (fig. 8.6).

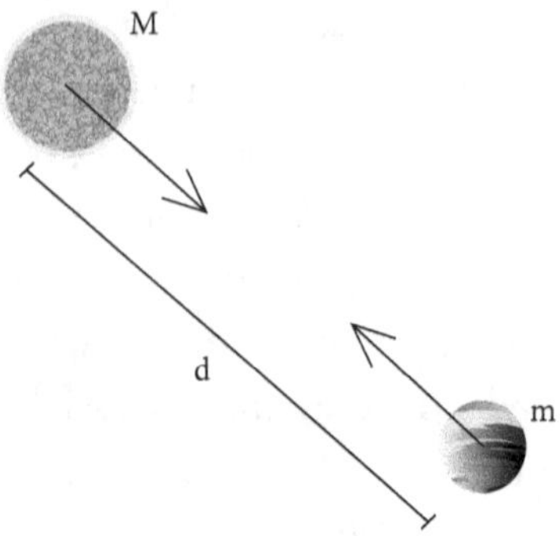

Fig. 8.6. The force of gravitation between two bodies is a mutual attraction acting along the line that joins the two bodies. It is directly proportional to the masses (M and m) and inversely proportional to the square of the distance (d).

He demonstrated with precise calculations and comparisons with the best and most recent astronomical data that all the planetary phenomena could be explained by this law of attraction. The solar system is a dynamic structure in which all the bodies attract one another. Since the mass of the Sun is much larger than that of the planets, however, in first approximation every planet forms a two-body system with the Sun. This is held together mainly by the mutual attraction of Sun and planet, the centre of mass of the system being very close to the centre of the Sun and the attractions of the other planets being negligible or reducible to minor perturbations. Kepler's laws descend as mathematical consequences from this arrangement. In the case of the Moon, Newton showed that, while the main force acting was the Moon-Earth attraction, the Moon-Sun attraction also had observable effects, which corresponded to the irregularities of the Moon's cycle that were known to astronomers as the lunar anomalies. He also showed that the tides were produced by the attraction of the Moon on the fluid mass of the oceans, with minor effects due to the Sun, and that the Moon also caused the slow conical motion of the Earth's axis that astronomers knew as the "precession of the equinoxes". He further concluded that comets also moved under the action of universal gravitation directed mainly toward the Sun. He calculated that a "great comet" observed in 1682 followed an elliptical orbit of large eccentricity, and predicted that it would complete a revolution in 75 years. The punctual reappearance of the comet was a sensational confirmation of his theory.[11]

[11]Another widely publicized confirmation arrived in 1736, when an expedition organized by the French Academy of Sciences measured the curvature of the Earth's surface near the Arctic Circle and concluded that the shape of the Earth

Newton refrained from using calculus in the *Principia* because he considered it to be still methodologically insecure. He applied instead Euclidean geometry, supplementing it with a geometric version of the calculus concepts. Instead of algebraically expressed fluxions, he used a method that he called "the method of first and last ratios of quantities". The method consisted of approximating the areas of curved figures by sums of polygonal areas, then letting the polygons become infinite and infinitely small. He proved than in the limit ("*in infinitum*") the polygonal areas became equal to the curved areas. For example, when proving that the area described by the orbital radius of a body moving under the action of a centripetal force is proportional to the time required to describe it (Kepler's law of areas), he imagined first a free body in rectilinear motion subjected at regular intervals to an instantaneous impulsive force—a quick tug, as it were—toward the centre. The path of the body would be a broken line, and the corresponding area a polygonal area composed of triangles (fig. 8.7).

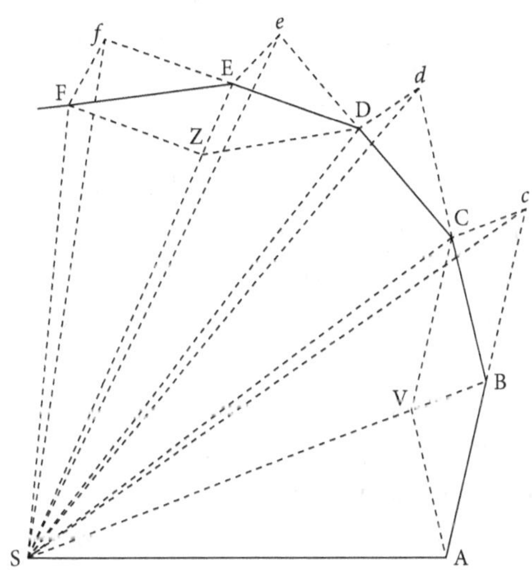

Fig. 8.7. Diagram of Proposition I, Book I of the *Principia*, illustrating the use of the method of last ratios to prove the law of areas. A body moves by inertia from A to B, from B to C, and so on, but receives instantaneous tugs toward S at regular intervals, at B, C, D, and F. In the limit of an infinite number of infinitely small intervals, the path ABCDEF becomes a curved line, and the acting force becomes a continuous centripetal force toward S.

He then completed the demonstration by letting the reader imagine that the triangles became infinite in number and infinitely small, so that the broken line became a curved line and "the centripetal force, by which the body is perpetually drawn back from the tangent of this curve, will act continually; and any described areas SADS, SAFS, which

is that of an "oblate spheroid" (a sphere-like solid flattened at the poles), as Newton had predicted as a combined effect of daily rotation around the Earth's axis and gravitational attraction toward the Earth's center.

are always proportional to the times of description, will, in this case also, be proportional to those times." (Newton, *Principia*, 1687, in Newton 1995, 239).

The range of universal gravitation was infinite. Newton's mechanics shattered the old idea of a finite ordered cosmos. Newton believed the Sun to be one among innumerable stars scattered uniformly in a space that stretched evenly in all directions. In this space, there was no predetermined centre and no special direction. Furthermore, he carefully showed that the celestial bodies moved as if they encountered no resistance from any surrounding medium. Celestial space was void of resistance. Newton did not exclude that space might be pervaded by a subtle from of matter ("aether"), invisible and intangible yet capable of somehow producing or simply carrying the gravitational attraction between bodies. He did not settle for the thought that gravitation was an action transmitted across empty distances. Mechanical philosophy admitted action by contact only, considering action at a distance to be an unintelligible notion. Nonetheless, Newton admitted that he was not able to find a mechanical explanation for gravitation, and declared that he preferred not to advance unsupported speculations about it. As he famously put it toward the end of the *Principia*, "I frame no hypotheses." (Newton, *Principia*, 1687, in Newton 1995, 342.)

For all his commitment to mathematical principles, Newton maintained a distinction between the "physical" and "mathematical" side of a theory. He did not offer his mechanics as a final and complete explanation of physical phenomena, but pointed out that in the *Principia* he aimed "only to give a mathematical notion of those forces, without considering their physical causes and seats." (Newton, *Principia*, 1687, in Newton 1995, 230.) Nevertheless, his followers did not pick up on his precautions. His rational mechanics and the theory of universal gravitation attracted increasing consensus and admiration in the course of the eighteenth century. His personal quirks, his unorthodox and sometimes fanatical theological beliefs, and his lifelong practice of alchemy were soon forgotten. For the thinkers of the Enlightenment, he became the beacon of reason, and his mathematical science became the paradigm of all science.

The principle of inertia implied the relativity of motion: rest and uniform rectilinear motion could be distinguished only relatively to a frame of reference conventionally taken to be at rest. Newton recognized that the relativity of motion was supported by common observations (once the Galilean method of stripping away the "accidental impediments" in real or thought experiments was accepted), yet on this point he was not willing to let empiricism rule. The concepts of space and time were for him beyond the limits of what can be learned from the senses. He asserted the need to distinguish between relative space and time and "absolute, true, and mathematical" space and time (Newton, *Principia*, 1687, in Newton 1995, 231):

1. Absolute, true, and mathematical time, of itself, and from its own nature, flows equably without relation to anything external, and by another name is called duration; relative, apparent, and common time is some sensible and external (whether accurate or unequable) measure of duration by the means of motion, which is commonly used instead of time, such as an hour, a day, a month, a year.

2. Absolute space, in its own nature, without relation to anything external, remains always similar and immovable. Relative space is some movable dimension or measure of the absolute spaces, which our senses determine by its position to bodies [...]

This implied also the distinction of absolute from relative motion, which Newton clarified using the Galilean example of the sailing ship (Newton, *Principia*, 1687, in Newton 1995, 232):

4. Absolute motion is the translation of a body from one absolute place into another; relative motion, the translation from one relative place into another. Thus in a ship under sail, the relative place of a body is that part of the ship which the body possesses, or that part of the cavity which the body fills and which therefore moves together with the ship; and relative rest is the continuance of the body in the same part of the ship, or of its cavity. But real, absolute rest is the continuance of the body in that same part of that immovable space, in which the ship itself, its cavity, and all that it contains, is moved. Wherefore, if the earth is really at rest, the body which relatively rests on the ship will really and absolutely move with the same velocity which the ship has on the earth. But if the earth also moves, the true and absolute motion of the body will arise partly from the motion of the earth in immovable space, and partly from the relative motions as well of the ship on the earth, as of the body in the ship [...].

Newton identified absolute space with the frame of reference in which the centre of mass of the solar system was at rest, and pointed out that the Sun moves slowly relative to absolute space. Therefore, at the same time as he was grounding Kepler's and Galileo's results in the new mechanics, he was moving beyond the geocentrism-heliocentrism debate.[12]

In Newtonian mechanics, space had become an invisible, empty, boundless container, in which any position was equivalent to any other and any direction equivalent to any other. It was no longer the textured environment in which we are immersed in everyday life, cluttered with sense experiences. It was the abstract, unlimited, and uniform space of Galilean thought experiments. In it, physical bodies were reduced to geometrical figures endowed with impenetrability (and, for the eighteenth-century Newtonians, also with the inherent power to attract other matter though empty space), and they moved without resistance along straight lines and geometrical curves. In short, physical space was identified with the abstract space of Euclidean geometry. The picture of the world was thoroughly mathematized.

Sources

Copernicus, Nicholas. *On the Revolutions of the Heavenly Spheres.* Translation and commentary by Edward Rosen. Baltimore and London: The Johns Hopkins University Press, 1992.

Dear, Peter. *Revolutionizing the Sciences. European Knowledge and Its Ambitions, 1500-1700.* Princeton, NJ: Princeton University Press, 2001.

Evans, James. *The History and Practice of Ancient Astronomy.* New York and Oxford: Oxford University Press, 1998.

Galilei, Galileo. *The Essential Galileo.* Edited and translated by Maurice A. Finocchiaro. Indianapolis and Cambridge: Hackett, 2008.

Kuhn, Thomas S. *The Copernican Revolution. Planetary Astronomy in the Development of Western Thought.* Cambridge, MA: Harvard University Press, 1957.

Newton, Isaac. in I. Bernhard Cohen and Richard S. Westfall, eds., *Newton. Texts, Background, Commentaries.* New York and London: Norton, 1995.

North, John. *Cosmos. An Illustrated History of Astronomy and Cosmology.* Chicago and London: The University of Chicago Press, 2008.

Westman, Robert S. "The Astronomer's Role in the Sixteenth Century: A Preliminary Study", *History of Science* 18 (1980), 105–147.

Westman, Robert S. *The Copernican Question. Prognostication, Skepticism, and Celestial Order.* Berkeley: University of California Press, 2011.

[12]Newton's belief in absolute time and space became the object of in-depth criticism in the last part of the nineteenth century by physicists such as Ernst Mach, Henri Poincaré, and Albert Einstein—a criticism that led to Einstein's creation of the theory of relativity.

Chapter 9

From the Mathematics of Chances to the Theory of Probability

9.1 The Mathematical Theory of Probability

The mathematical theory of probability emerged from the confluence of two currents of studies of the seventeenth century, the calculation of chances and the analysis of rational belief. The first was, initially, a peculiar niche of problem-solving where mathematically talented authors could showcase their skills by answering questions of odds, advantage, and fairness in games of chance, such as dice or cards. The second was a philosophical endeavour concerned with the possibility of knowing under conditions of uncertainty. This was a widespread concern at a time in which many thinkers felt that the old philosophical systems were crumbling, and a deep mistrust in the possibility of human knowledge threatened to settle in their place. Before the seventeenth century, chance was regarded as an aspect of reality that was unknowable and uncontrollable. Devices that produced random displays, as for example the ancient *astragali* or knucklebones, and later dice and cards, were used for entertainment and gambling—as eloquently attested by the existence of anti-gambling laws in many cultures since Antiquity. They were also used for divination, as unpredictable phenomena were sometimes interpreted as the expression of the inscrutable will of the gods. Nevertheless, they remained outside the scope of rational inquiry and, hence, were excluded from mathematical study. There were, of course, occasional attempts to enumerate the possible outcomes of chance displays, as well as empirical rules of fair betting and judicious playing. Nevertheless, no systematic calculation was carried out and no attempt at formulating a theory was made. The aim of science (*episteme* or *scientia*) was certain knowledge. The method of science was demonstrative logic, or deduction, the kind of reasoning that proceeded from accepted premises to necessary conclusions and absolute certainty. If affected by uncertainty, knowledge was nothing but opinion, and an opinion could be more or less "probable", which meant credible on the basis of argument. Arguments were of various kinds, ranging from approval by authority in medieval and Counter-Reformation Jesuit theology, through support by testimony and forensic evidence in jurisprudence, to collection of experimental evidence in modern natural philosophy. For example, when Andreas Osiander advised in his foreword to the first edition of *On the Revolutions of the Heavenly Spheres* that Copernicus's hypotheses needed "not be true or even probable", he was dismissing Copernicus's claim that ancient authors had supported the motion of the Earth. Similarly, Galileo argued that although the heliocentric hypothesis was "improbable", correct reasoning and telescope observations showed that it was in fact credible, whereas the Inquisition accused him precisely of having defended the "Copernican opinion" as probable, even though it was contradicted by the Holy Scripture.

Around the mid-seventeenth century, a number of new developments in mathematics, logic, and legal and financial thinking concurred to create a new field of study. The calculus of chances was first invented to solve some nagging problems related to gambling, and then was brought to bear on an increasing spectrum of other activities in which unpredictable events were involved. Annuity and insurance contracts, rates of births and deaths in a population, errors of measurement, and the optimization of judicial sentencing were among the first subjects. At the same time, the meaning of probability shifted away from approbation by authority and toward the amount and force of evidence supporting a given hypothesis, or the degree of rational belief. The mathematics of chances was put to work to evaluate this new kind of scientific knowledge, uncertain yet quantifiable, giving rise to the modern, quantitative concept of probability and to its mathematical theory. This new study, the mathematical theory of probability, was classified by the philosophers of the Enlightenment as one of the "mixed mathematics" alongside physico-mathematical fields like astronomy, optics, and mechanics. It became a tool for managing uncertainties and making decisions in an ever growing variety of fields. In the words of the authors of a book on the history of this subject, probability theory was the result of two movements, "a new pragmatic rationality that abandoned the traditional ideals of certainty; and a sustained and remarkably fruitful attempt to apply mathematics to new domains of experience." (Gigerenzer et al. 1989, 1.) Eventually, after having proved its worth in areas as diverse as the analysis of experimental observations, molecular physics, and the study of heredity, in the early twentieth century the theory of probability became a branch of "pure mathematics", with limitless possibilities of being applied but in itself abstract and independent of any of its applications.

9.2 The Calculus of Chances

The Mathematics of Gambling and the Principle of Equally Likely Outcomes

Most histories of mathematics locate the birth of the mathematics of chances in an exchange of letters that occurred in 1654 between two Frenchmen, the philosopher and mathematician Blaise Pascal and the lawyer and mathematician Pierre de Fermat. Although Pascal and Fermat did not know each other personally, they belonged to a circle of intellectuals who placed a great value on mathematics. They believed mathematical reasoning to be not only the key to understand the physical world, but also the antidote to what they identified as the two intellectual plagues of their time, dogmatic sectarianism at one extreme and philosophical skepticism at the other extreme. In their letters Pascal and Fermat discussed gambling problems that had been presented to Pascal by a friend, the Chevalier de Méré. With their solutions, they laid out the foundations of the calculus of chances.

The first of de Méré's questions concerned the correct way to calculate the odds of chance outcomes that are compounded of simpler outcomes. De Méré wanted to know, specifically, how many throws of two dice will give favourable odds of casting two sixes. Apparently, he had made some calculation according to which the minimum number would be 24 throws. But he knew that gamblers knew that it was disadvantageous to bet on a double six in less than 25 throws. Did this mean that rigorous mathematics was inapplicable to chance? Pascal solved the problem—and thereby demonstrated that mathematics could conquer even that monster, chance—by showing that the correct calculation did in fact agree with the empirical rule. First of all, he clarified that it was easier to operate with *chances* rather than odds (to clarify, the odds of throwing a six with a single die are 1 to 5, the chances are 1/6). Second, he established that the chances of an outcome were to be calculated as the *ratio of the number of ways in which the outcome could be produced over the number of all possible ways to produce an outcome.* It was not new to define chances in this manner, but the fundamental precondition that both Pascal and Fermat made clear was that the "ways to produce an outcome" which were to be counted had to be *equally likely* cases. This precondition, now unimaginatively known as the "principle of equally likely outcomes", had been already set out a few times by scattered authors in the course of the previous three centuries. Yet, it was only with the Pascal-Fermat correspondence that it took definitive root and grew into a reliable technique to calculate chances and odds.

Some medieval texts on divination contained attempts to enumerate the possible outcomes of a chance setup; thus, they alleged that there are 21 ways of casting two dice. This number is correct only if the outcomes are counted without regard to the order of the dice. In this count, for example, the configuration in which the first die shows a two and the second shows a three is considered to be the same outcome as the configuration in which the first die shows a three and the second shows a two (fig. 9.1).

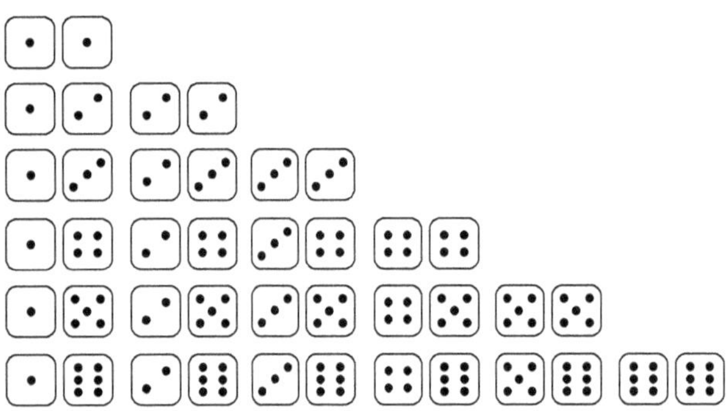

Fig. 9.1. One way of counting the casts two dice. In this count, the outcomes are not equally likely.

This way of counting outcomes suited the purpose of divination, since predictions were read from the display independently of the odds. In gambling, however, the odds of the different outcomes were of concern. Gamblers knew that the 21 displays of two dice would not occur with the same facility. They knew, for instance, that the odds of casting a five were higher than the odds of casting a four. The earliest known author to explain mathematically why it was so was Girolamo Cardano, the sixteenth-century doctor, astrologer, and abacist who is famous for publishing the solution of cubic and quartic equations in his algebra treatise, *Ars Magna*. Besides being a prolific author and shrewd self-promoter, Cardano was a gambler. Among his many unpublished works, he left a collection of writings on gambling in which he dispensed advice on how to play fairly and how to spot cheating, mixed with autobiographical, moral, and legal considerations. As an example of his style, he warned, probably speaking from personal experience, that doctors, lawyers, and other professionals might suffer a loss of reputation if they gambled, in the same way as they would if they practised music. He offered as a "Fundamental Principle of Gambling" a principle of equality, upon which he seemed to want to build an argument that gambling did not have to be disreputable (quoted in Ore 1965, 189):

> The most fundamental principle of all in gambling is simply equal conditions … of money, of situation … and of the dice itself. To the extent to which you depart from that equality, if it is in your opponent's favour, you are a fool, and if in your own, you are unjust.

Then he explained that, in order to be fair, a wager had to be proportional to the number of ways a favourable outcome could be produced, and that the "ways to produce an outcome" to be counted had to be equally likely events. For example, it is not equally likely to cast two and two as to cast one and three, because the latter can be obtained with one on the first die and three on the second as well as with three on the first and one on the second. This means that there are three equally likely ways to cast a four, while there are four equally likely ways to cast a five (fig. 9.2).[1]

[1]It is possible that the source of Cardano's idea was a thirteenth-century Latin poem, printed and circulated during the Renaissance, titled *De vetula*. Combinatorics, the study of enumeration and groupings of discrete elements, is a branch of mathematics with a long history. In modern combinatorial language, configurations in which only the numbers of elements in subgroups are known but not which elements is in which subgroup are called partitions, and configurations in which it is known which element is in which subgroup are called permutations. A configuration of two dice in which one die displays one number, for example 1, and one die displays another number, for example 2, regardless of the order of the dice, is a partition. This partition can be produced by two distinct permutations, one in which the first die displays 1 and the second die displays 2, and the other in which the first die displays 2 and the second 1.

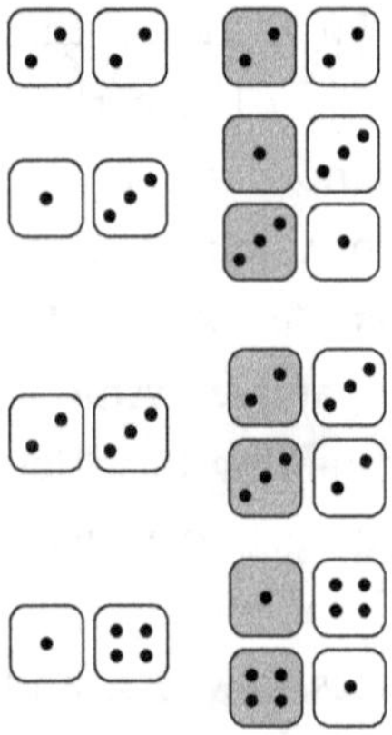

Fig. 9.2. Configurations of two dice that correspond to a four and to a five. The configurations on the left are not equally likely to occur; those on the right are (provided that both dice are unbiased and independent.)

Cardano called the complete set of equally likely outcomes of a game the "circuit". He also understood that, since the circuit of one die contained 6 outcomes, the circuit of two and three dice contained respectively $6\times6=36$ and $6\times6\times6=216$ outcomes, and that the correct way to compute the chances in these cases was to multiply the chances of the individual dice (fig. 9.3).

Cardano's meditations on gambling were found among his numerous manuscripts after his death; they were put together and published only in the following century, nine years after Pascal and Fermat's exchange, with the title *Liber de ludo aleae* (*Book on the Game of the Die*). No one can say for sure what would have been the fate of his ideas if they had been made public in the sixteenth century. In the early seventeenth century,

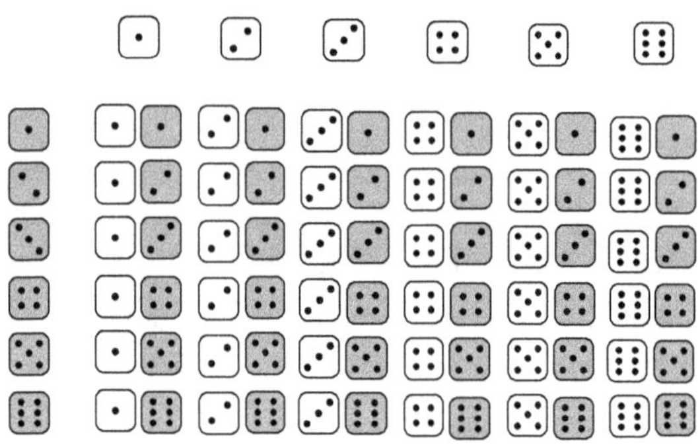

Fig. 9.3. When casting one die, there are 6 equally likely outcomes, and the chance of each of them is 1/6. With two dice, there are 36 equally likely outcomes, and the chance of each of them is 1/36.

Galileo was asked a question about the odds of dice casts, and he wrote an essay in which he provided a count of equally likely outcomes. But there is no evidence that this work stirred enough interest to prompt more work on the matter. When Pascal and Fermat re-discovered the principle of equally likely outcomes in the mid seventeenth century, they did so less with a personal interest in games and more with an eye to the universal power of mathematics.[2] Catching the rising wind of mathematical rationalism, their ideas soon sailed out of the idiosyncratic world of gambling—even though the games of chance remained the paradigmatic examples on which, to this day, people learn about probability.

The modern theory of probability is an axiomatic theory, and the principle of equally likely outcomes is one of its fundamental axioms.[3] In every application of the theory, the calculation of probabilities starts with an assignment of equal probabilities to a set of elementary events. Today's authors recognize that it is not possible to define probabilities without defining a set of equally likely elementary events. At the same time, it is not possible to provide a formal definition of "equal likelihood" without referring to equal chances or probabilities. A formal definition of chance or probability seems therefore destined to be hopelessly circular. As is the case for modern axiomatic geometry, most mathematicians prefer not to give an explicit definition of probability but to let it be implicitly defined by the axioms: probabilities are the numbers that satisfy the probability axioms. The axioms fully determine how compound probabilities are to be calculated once the equiprobable set of elementary events ("sample space") is provided, without any reference to what probability is, if it is anything, in the real world. Philosophers, on the other hand, are still debating the meaning of probability in the real world, with different schools of thought fiercely defending their definitions against the adversaries. To simplify a subtle and still fluid set of issues, there are three main schools of thought: the frequentist school, according to which probability is the relative frequency of occurrence of an event in the long run; the so-called subjectivist or Bayesian schools, according to which probability is the degree of rational belief; and the propensity school, according to which probability is the intrinsic propensity or facility with which an event will occur. The different interpretations of probability share the intuitive notion of equally likely (or "equiprobable") outcomes, which in turn derives from the idealized experience of symmetrical devices or setups, such as a perfectly cubic die made of perfectly homogeneous material, a perfectly balanced coin tossed in a perfectly balanced manner, a set of perfectly equal-sized and shaped cards, and so on. In principle, for idealized artifacts of this kind, the laws of physics do not favour any of the possible outcomes over the others. These objects, then, can

[2] Then as now, it seems that mathematicians took and interest in gambling problems for the challenges that they presented and, at least in part, for the pleasure of correcting the fallacies of less mathematical minds.

[3] The axiomatic theory of probability was formulated in 1933 by the Russian mathematician Andrey Kolmogorov.

be regarded as the conceptual seeds of the old definition of chances as the ratio of favourable outcomes to total number of outcomes, and hence of the different definitions of probability that derived from it.

Once a set of equally likely elementary events is established, it is possible to calculate the chances of more complex outcomes. Let us return to the first of de Méré's problems, which concerned the casting of a double six with a pair of dice. Pascal and Fermat, like Cardano, understood that when the outcome under consideration consists of the simultaneous occurrence of two independent elementary events, the chances of the independent events have to be multiplied. Casting a double six with a pair of dice is obtained with a six on one die and a six on the other. Hence, the chances of a double six in a single throw are $\frac{1}{6}\cdot\frac{1}{6}=\frac{1}{36}$ (fig. 9.3).[4] Similarly, since the chances of not throwing a double six in a single throw are 35/36, the chances of not throwing a double six in n throws are $\left(\frac{35}{36}\right)^n$. From this, Pascal calculated that $n = 25$ was the minimum number of throws in which the chances of not throwing a double six were less than 1/2.

The Problem of Points, or on Sharing Future Profits Fairly

The second of de Méré's questions was the problem of how to divide the prize among two players in a game of chance when the game was interrupted before its completion. This problem was known as the "division problem" or "problem of points", and it had been discussed since the fourteenth century in handbooks of practical mathematics. Although concerned with games, it was a problem analogous to the business problems that were the breeding ground of commercial and legal arithmetic. It was similar, for example, to some questions of business partnerships and

[4] This is an "AND" combination. A compound outcome can also be produced by an "OR" combination, that is, by the separate occurrence of simpler events. In this case, it is necessary to distinguish between mutually exclusive and non-exclusive events. For mutually exclusive events, the chances have to be added. For example, throwing an even number with one die is obtained by throwing a two OR a four OR a six. These events are mutually exclusive, and each of them has 1/6 chances to occur. Hence, the chances of an even number are 3/6 = 1/2. For an OR combination of non-exclusive events, the chances of the combination are equal to the sum of the simple events minus the chances of the simultaneous occurrence of the events. For example, throwing a six with two dice is compounded of throwing a six with the first die OR with the second, but since the two sixes are not mutually exclusive, the sum of their separate chances includes the chances of a double six twice. Hence the chances of throwing a six with two dice are $\frac{1}{6}+\frac{1}{6}-\frac{1}{6}\cdot\frac{1}{6}=\frac{11}{36}$.

inheritance in case of death of one of the partners that al-Khwarîzmî discussed in his *Book of Reunion and Reduction*. The abbacists tried applying algebra to it, but without success. An example given by Luca Pacioli in the *Summa de arithmetica* (1494) was as follows:

> A team plays ball so that a total of 60 points is required to win the game, and each inning counts for 10 points. The stakes are 10 ducats. By some incident, they cannot finish the game and one side has 50 points, and the other 30. What share of the prize money belongs to each side?

Although Pacioli noted that opinions differed on the correct solution and that every argument seemed to be insufficient, he opined that the right answer was to divide the prize in the proportion of the points already scored. Fifty years later, Tartaglia pointed out that Pacioli could not be right, because either player could still have won the whole prize had the game been completed. Tartaglia commented that "the resolution of such a question must be judiciary rather than mathematical," for in whatever way the division was made there might be cause for litigation. In his view, there was little substance in such facetious questions, which some people liked to bring up just to rouse an argument. Cardano, too, pointed out Pacioli's mistake; he hinted that the correct solution was to divide the stake according to the chances of each player to win, but did not develop the complete solution. The question continued to stir interest and remained unsettled until Pascal and Fermat found a definitive answer. Fermat and Pascal (like Cardano again, though they did not know) agreed that the division should be made in proportion to the chances that each player had to win if the game were brought to completion. Arguably, the acceptance of this kind of solution relied on a shared economic mindset. By this time, financial investments had become commonplace, and it had become common knowledge that the value of one's participation in a business partnership depended on the potential for future profits. Fermat found the solution in a specific case, enumerating the equally likely outcomes. Pascal worked out a general method of solution that would be more efficient in the case of many equally likely outcomes. He started with the simplest case, the case in which both players, let us call them A and B, need only one more round of the game to win, and hence have each 1/2 chances to take the whole prize. In the case in which A needs one more round and B needs two, A will have 1/2 chances to score the next point and win, and 1/2 chances not to score the next point, in which case she and B will again be even; in total, then, A will have $\frac{1}{2}+\frac{1}{2}\cdot\frac{1}{2}=\frac{3}{4}$ chances to win, and B will have $\frac{1}{2}\cdot\frac{1}{2}=\frac{1}{4}$ chances (figs. 9.4a, 9.4b).

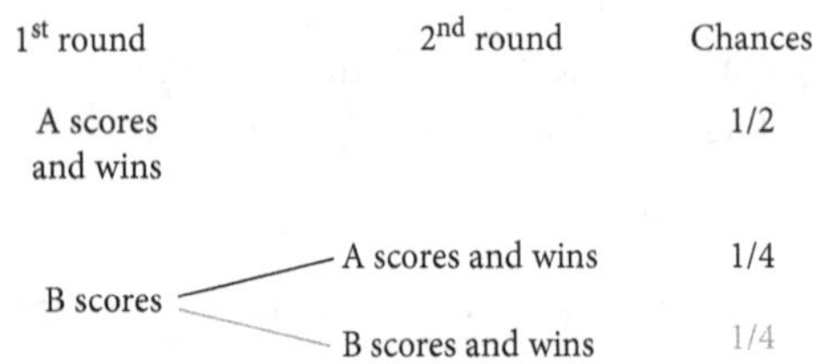

Fig. 9.4a. Possible outcomes and their chances when player A needs one more point and player B needs two more points to win.

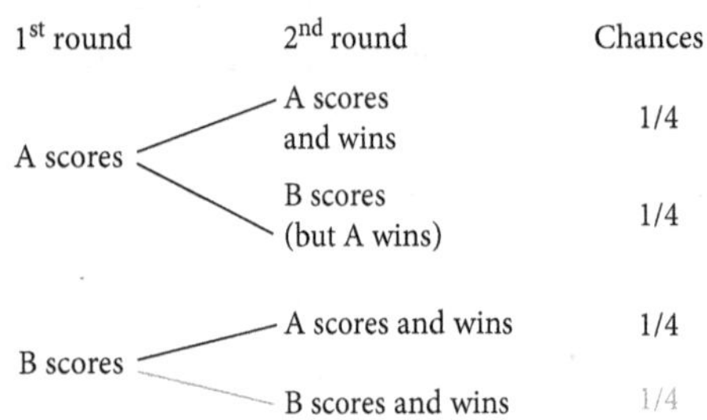

Fig. 9.4b. Enumeration of equally likely outcomes when A needs one point and B needs two points to win.

The prize is therefore to be divided in the proportion of 3 to 1. Reasoning step by step to more complex cases, at each step using the results found in the previous step, Pascal found that the general solution involved an array of numbers that he called the "arithmetic triangle" and is now known as "Pascal's triangle" (fig. 9.5)[5]:

[5] Pascal went on to write (among other works in hydrostatic, projective geometry, and philosophy) a *Treatise on the Arithmetic Triangle.* The terms in Pascal's triangle are also known as binomial coefficients, as they are the coefficients of the expansion of the powers of a binomial, e.g. $(a+b)^n$ when the exponent, n, is a positive integer. For example, the coefficient of the 4th power of the binomial are the terms in the 4th row of the triangle, $(a+b)^4 = a^4 + 4a^3b + 6a^2b^2 + 4ab^3 + b^4$. The binomial coefficients were known in China, India, and Islam before they appeared in Europe in the work of Pascal, Newton, Jakob Bernoulli, and others. Their history is long and complex, and still far from clear. Pascal also gave a description of his step-by-step method which was the first explicit formulation of a general method of proof that is called "mathematical induction". Mathematical induction is a special technique of demonstration in deductive mathematics. It is not to be confused with the method of induction, a method of scientific research that aims as formulating general theoretical laws or axioms from particular observations or experiments, or with inductive logic, the type of reasoning that aims at making inferences from premises without certainty (see below).

1						
1	1					
1	2	1				
1	3	3	1			
1	4	6	4	1		
1	5	10	10	5	1	
1	6	15	20	15	6	1
...						

Fig. 9.5. Pascal's triangle. Starting from 1, the successive terms of each row can be calculated by adding the two previous terms in the previous row.

In general, if A needs n_A more rounds and B needs n_B more rounds to win, then the maximum number of rounds needed to complete the game is $n_A - 1 + n_B - 1 + 1 = n_A + n_B - 1$. The total number of equally likely outcomes is, therefore, $2^{n_A + n_B - 1}$. Pascal showed that the number of outcomes favourable to A was equal to the sum of the first n_B terms in the $(n_A + n_B)^{th}$ row of the triangle, and the number of outcomes favourable to *B* was the sum of the remaining n_A terms in the same row. For example, when $n_A = 1$ and $n_B = 2$, the number of outcomes favourable to A is $1 + 2 = 3$, the number of outcomes favourable to B is 1, the total number of outcomes is 4, and the chances to win for *A* and *B* are respectively 3/4 and 1/4. When $n_A = 2$ and $n_B = 3$, the numbers of outcomes favourable to A and B respectively are $1 + 4 + 6 = 11$ and $4 + 1 = 5$, the total number of equally likely cases is 16, and the respective chances to win are 11/16 and 5/16. And so on (fig. 9.6).

To return to Pacioli's problem, in that case $n_A = 1$ and $n_B = 3$; therefore, the first team has 7/8 chances to win, and the prize is to be divided in the proportion of 7:1.

Uncertainty, Expectations, Decisions: A New Pragmatic Rationality

Shortly after his correspondence with Fermat, Pascal underwent a religious conversion and became a member of the Jansenist sect. The Jansenists were Catholic fundamentalists who spurned science as a distraction from penitence and prayer. In a later book on theology, however, Pascal applied his grasp of chances to the problem of religious belief. Addressing those who thought that reason was powerless to decide whether God existed or not, he compared the decision to believe to a wager. You wager that God exists (evidently, Pascal meant the God of the Jansenists). If He does exist, you win the bet and the prize is the salvation of the soul. If He does not exist, you lose the bet, but the price to pay is something of finite value, for it is, at worst, your human life (because you would waste it living like a Jansenist, presumably.)

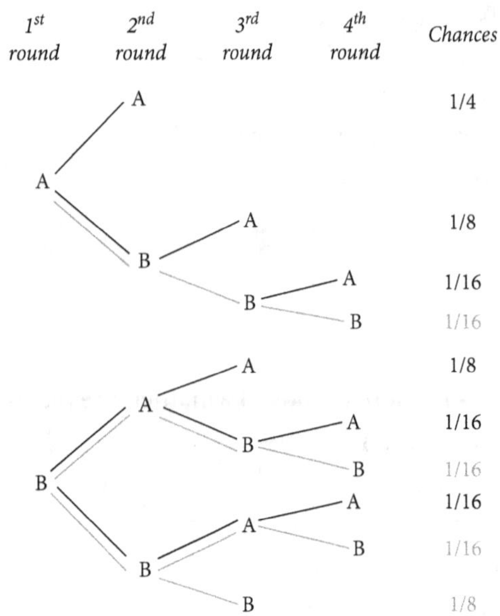

Fig. 9.6. Outcomes favourable to A and B, and the respective chances, when A needs 2 more points and B needs 3 more points to win.

The chances of winning or losing are unknown. Yet, Pascal argued, no matter how low the chances may be, they are not zero, and this means that it is always worthwhile to wager "an infinite life of infinite happiness" against one finite life. This quirky argument is famous as "Pascal's Wager". One historian commented,

> Much criticism has been expressed against Pascal's logic, and one must admit that this is hardly a field for applied mathematics. But even Pascal himself cannot have believed that his argument should carry a convincing mathematical weight; however, if his design was to create a thought-provoking parable, he succeeded admirably. (Ore 1960, 417.)

Yet, it was not Pascal's *logic* that was weak, but his premises. Pascal assumed that it made sense to treat religious belief as a pragmatic decision, and he equated pragmatic decisions to betting. He also assumed that the binary choice, "either *this* God exists or there is no God" exhausted the possible outcomes. Many would disagree with these premises.[6]

[6] One criticism of Pascal's Wager is that you cannot "decide" to believe in God; religious conversion is not a matter of deliberation. Apparently, however, the Jansenists were convinced that if you act piously *as if* you believed in God (and that is something you can decide to do) you would eventually end up believing. Another criticism is that to be religious for self-interest defeats the purpose of being religious. One more criticism is that it is possible that some god or gods exist, but not the god of the Jansenists. If that turns out to be the case, the retribution for believing in the wrong god is not likely to be eternal bliss.

Nevertheless, if they are granted, the structure of the argument is a valid example of the new kind of logic that was taking form at the time, and that today's logicians call "inductive logic". In inductive logic, the inferences do not follow from the premises with iron necessity, as is the case in deduction, but the evaluation of chances is used as a means to make judgements and decisions. For our purposes, Pascal's Wager illustrates how chance, once it was forged into a mathematical tool, became—at least in the minds of the promoters—a vehicle for mathematics to colonize continents of experience that were oceans apart from its traditional subjects. The calculation of chances became the tool of a new pragmatic rationality, a method for decision making with seemingly limitless potential of application.

Pascal's Wager also illustrates a central principle of the new logic. When a choice between potential outcomes has to be made it is not sufficient to compare only the payoffs or only the chances. Chances and payoffs ought to be weighed together. This principle was formulated mathematically by the Dutch physico-mathematician Christiaan Huygens in his book, *De ratiociniis in ludo aleae* (*On Reasoning About the Game of Dice*, 1657), the first published book on the calculus of chances. Upon learning by hearsay of Pascal's and Fermat's results, Huygens developed the idea that the relevant quantity to make decisions in a game, including the decision of whether to enter the game or not, is the product of the chances by the payoff of the outcome under consideration. He called this quantity "the value of a chance"; today, it is called "mathematical expectation" or "expected value". A cluster of theories that branch from applied mathematics into psychology, philosophy, economics, military strategy, and more, and go under the umbrella name of "decision theory" have grown around the concept of mathematical expectation. The core principle of decision theory is that the rational strategy to make a decision is to consider all the possible outcomes of the decision, calculate the chances and the "utility" (the payoff) of each outcome, which may be positive or negative, and then calculate the expected value of the decision by summing up the products of the chances by the utility of every outcome. Then, the most rational decision will be the one that has the highest expected value.[7]

Huygens explained the concept of expectation by making the example of someone hiding three coins in one hand and seven in the other, and letting you have the money that is in the hand you choose. How much is it fair for you to pay to play this game? This game has the

[7] Pascal's Wager followed this principle. If you were willing to grant Pascal's premises, the decision to live your life as if you believed in God had the maximum expected value independently of what the chances of God's existence were, as long as they were not zero.

expected value of five coins to you because if you pay five coins to play, then you will have equal chances of winning or losing the same amount, and so will the person with whom you are playing. Although Huygens's explanation sounds somewhat abstruse today, it makes clear that the definition of expected value was rooted in the view that a game was like a business contract. As such, it had to be evaluated according to an ethics of fairness, that is, of equality for all the participants. Put in other terms, a game will be fair if the *total* expected value for each participant is zero. As another example, let us consider the game of roulette as played in Monte Carlo. The roulette has 18 red numbers, 18 black numbers, and 1 zero, and the ball is equally likely to stop in any one of the 37 slots of the wheel. The prize for a "straight up", or single-number bet, is 36 times the wager. Thus, if you bet $1\$$ on a number, your total expected value is $\frac{1}{37}35\$+\frac{36}{37}(-1\$)=-0.027\$$. This value is called the "house edge", since it is the amount that a player is expected to lose, and the house to gain, on average. Would you play this game? The decision that maximizes the mathematical expectation is not to play—unless, of course, the thrill of the game is a positive value that you are willing to include in the calculation. Consider lotteries. From the standpoint of mathematical expectation, buying a lottery ticket is always an irrational decision, because no matter how rich the prize and how cheap the ticket, the chances of winning are so low that the expected value is always negative for the buyer. A lottery is, in fact, a fund raiser for the state or organization that issues it. The French statist Talleyrand called the state lottery "a tax on the poor". Others, less charitably, call it "a tax on the stupid". Nevertheless, people buying a lottery ticket are buying a dream of changing their lives, and they are probably comparing the calculably low chances of winning the lottery to the incalculably lower chances of becoming rich in any other way. It is not always easy to quantify all the utilities. Moreover, except in games where the setups are symmetrical and the rules are defined in advance, it is almost never simple to calculate the chances.

9.3 The Classical Theory of Probability

The *Art of Conjecturing,* Probability, and the Law of Large Numbers

Cardano, Pascal, Fermat, and their early followers did not use the term "probability" in the mathematics of gambling. The term did not yet belong in mathematics. It belonged in legal, philosophical, and theological debates, and in empirical disciplines, such as medicine, alchemy, and the rising experimental studies of chemistry, pneumatics, electricity, and so on. The advent of experimentalism shifted the weight of probability from the word of illustrious authors of the past to the direct witnessing of modern experimental demonstrations. At the same time, it gave new urgency to the question of what kind of argument made a hypothesis more probable than another. Writing about hypotheses on the physics of light, for example, Huygens argued that experiments could provide "proofs of probability", by which he meant that they could attain "a degree of probability which very often is scarcely less than complete proof." Yet, neither did he attempt to apply his results in the mathematics of gambling to quantify the probability of a physical hypothesis, nor did he use the word probability in his treatise on games of chance. For Huygens as for Galileo, chances and probability were two different concepts.

The fusion of chance and probability was pioneered by the Swiss mathematician Jakob Bernoulli in an unfinished book titled *Ars conjectandi* (*The Art of Conjecturing*), which was published posthumously in 1713. The title of this book was a nod to the most influential textbook on logic of the time, *Ars cogitandi* (*The Art of Thinking*), also known as *Port Royal Logic*, which was published in 1662 by two French philosophers of the Jansenist group, probably with some input also from Pascal, and which contained a discussion of probability similar in spirit to Bernoulli's work. In his book, Bernoulli first reported Huygens's calculus of chances, adding extensive commentary and new problems and calculations. Then, he set out to extend the new mathematical method from games to less frivolous problems of uncertainty in "civil, moral, and economic matters." He argued that scientific knowledge did not need to coincide with total and absolute belief. Partial belief could be made scientific because it could be quantified by means of an analogy between probability and chance. His aim was to mathematize an existing tradition of evaluating the probability of proofs in legal trials by applying to it the calculation of chances. He defined probability as the "degree of rational belief" or, equivalently, the "degree of certainty". As he put it, "Probability is a degree of certainty, and differs from certainty as a part from a whole." Thus, probability became mathematical. Probability became a number between 0 and 1, the value 0 corresponding to absence of belief, the value 1 to absolute certainty, and the value 0.5 to complete uncertainty. A value of probability could be assigned to a hypothesis on the basis of the evidence available in the same way as the share of a prize in an incomplete game was allotted according to the chances to win. Bernoulli stressed that practical decisions did not necessarily demand perfect certainty. A sufficiently high probability, as for example 0.99 or 0.999, would be indistinguishable from certainty for all practical purposes. Deploying the legal vocabulary, he called this sufficiently high probability "moral certainty". Bernoulli's definition of probability was quickly adopted by other mathematicians, giving rise to the mathematical theory of probability. Probability understood as degree of belief characterized the first phase of this theory, the so-called classical theory of probability, which came to an end shortly after the publication of its most comprehensive and influential exposition, *The Analytical Theory of Probability* by the Frenchman Pierre-Simon de Laplace (1813).

Degree of belief is an *epistemic* quantity, that is, it is the quantitative expression of a mental state of the knowing subject, whereas chance is a property of the objects to be known. Starting from the 1820s, theorists of probability rejected the epistemic definition of probability; they branded it as a "subjective interpretation" and distinguished it from their own "objective interpretation", according to which probability is an attribute of the objects of knowledge and therefore independent of anyone's knowledge or belief. For the classical probabilists from Jakob Bernoulli to Laplace, however, the fact that probability was epistemic did not mean that it was "subjective" in the sense of personal and dependent on private perceptions. They believed in a single universal rationality, one correct way of thinking that was shared by all rational men. Every rational man, presented with the same evidence, would come to the same conclusion. Hence, anybody whose mind was not obfuscated by ignorance, error, and superstition would assign the same probabilities. Furthermore, the beliefs of rational men, if acquired from correct experience, would reflect the truth about the world. Correctly assigned probabilities would then correspond to properties of external reality, and were thereby objective even though epistemic.

Bernoulli's goal was to use the mathematics of chances to calculate probabilities. In games of chance, the chances of every outcome could be calculated prior to playing the game, just by knowing the setup and rules of the game. In other terms, chances were knowable a priori, that is, before any observation was made. The analogy between chances and probability presupposed that probability too was in principle calculable as the ratio of favourable to total outcomes. Bernoulli realized, however, that the a priori method of calculation was not helpful in real life situations because most often it was impossible to enumerate the equally likely cases. Who could determine in advance how many diseases could affect the body, and how more likely would a disease be fatal than another? Who could count the possible changes in the weather? He found a way out of this difficulty in the common sense notion that observations could supply information about the unknown probabilities, and that with more and more observations the information would become increasingly accurate. Cardano had already expressed the nugget of this notion. He asserted that the circuit of a game corresponded to the minimum number of rounds of the game necessary to produce all the outcomes. Thus, when casting a die, the circuit consisted of six rolls. But because of randomness, in practice some of the outcomes would be repeated, and therefore the number of rolls actually needed to cast all six outcomes would always be higher than six. Still, the random fluctuations could be evened out by repeating the rolls, so that in a sufficiently long series each outcome would be produced approximately 1/6 of the times. For the same reason, when tossing a coin many times, assuming that the coin is not biased, one expects heads to be cast approximately as many times as tails, even thought the outcome of every single toss is entirely unpredictable, and it is in fact well possible to have "lucky" or "unlucky" runs in which heads or tails dominate. Cardano was a man of his time, the Renaissance. He believed that luck was a real power existing in the world, and warned players about it.

Jakob Bernoulli belonged to a different time: he was a Newtonian. In open contrast with what he considered old superstitions, he believed the world to be governed by mathematical laws, not occult powers. In his view, the behaviour of coins, dice, and the like was in principle as regular and predictable as that of the planets, because it was determined in a necessary way by the laws of mechanics and by knowable forces such as gravity. No real randomness existed in the world of the Newtonians. Theirs was a *deterministic* vision: the universe was a vast machine, a wondrous and exquisite clockwork clicking away in a predetermined manner. It was only the limited human capacity of apprehending the relevant physical variables (initial positions and velocities of every particle of matter, exact masses, exact effects of frictions and resistances, and so on) that created uncertainties and made the use of probability necessary. From the standpoint of his deterministic worldview, Bernoulli assumed that every uncertain occurrence was exactly analogous to a game having a fixed, albeit unknown, ratio of favourable to possible outcomes. For instance, he maintained that the body is like a "tinder box" containing a predetermined number of latent diseases with predetermined chances of manifesting themselves, in the same way as lottery numbers are drawn according to their fixed ratios. He saw the effects of chance as small irregularities that would be gradually smoothed over a large number of observations, eventually producing predictable effects.

In repeated trials of a chance process, the *relative frequency* of an outcome is the ratio between the number of occurrences of the outcome and the total number of trials.

Symmetrical chance devices, in which the chances of each possible outcome are constant from one trial to the next, have a tendency to produce stable frequencies in the long run, and the relative frequencies tend to be close to the chances. Therefore, Bernoulli reasoned that long-term frequencies could be used to estimate a posteriori—that is, after the observations, empirically—probabilities that are unknown a priori. The question was, how many observations were necessary to eliminate the disturbing effects of chance fluctuations? According to Bernoulli (Bernoulli 2006, 328),

> To give an example of what I have in mind, suppose that there are hidden in an urn, unknown to you, three thousand white tokens and two thousand black, and that in order to investigate their number by experiments, you take out one token after another (but each time putting back the one that you have taken before you choose the following one, lest the number of tokens in the urn be diminished) and you observe how many times a white token comes out and how many times a black one. It is asked whether you can do this so many times that it becomes ten, a hundred, a thousand, etc. times more probable (that is, that in the end it becomes morally certain) that the number of times in which you have chosen a white and in which you have chosen a black will have to each other the same ratio of three to two that the numbers of tokens secretly enjoy than some other different ratio.

Bernoulli's major achievement was to transform the common-sense intuition about stable long-term frequencies into a theorem. Identifying probability and chance and using the rules of the calculus of chances as axioms, he derived the relation between the probability of an outcome in a single trial and the probabilities of the relative frequencies in many trials. He called it his "golden theorem"; it would later become famous as the Law of Large Numbers.[8]

Bernoulli considered a chance setup in which there were two possible outcomes, generically termed "success" and "failure", having constant probabilities. Occurrences of this kind, in which a process is repeated multiple times under the same conditions, are known as "Bernoulli trials". The simplest example is a series of coin tosses or, equivalently, a simultaneous toss of many equal coins. The probability of either heads or tails in a single toss is 1/2, and it remains the same at every toss: the outcome of any toss is not affected by

[8] Stricter versions of this law were formulated in the nineteenth century. We shall only consider Bernoulli's version.

that of the previous. The tosses are therefore Bernoulli trials with probability of success (either one of the two outcomes) equal to 1/2. Combining the probabilities according to the calculus of chances, it is possible to compute the probability of any frequency of successes. For example, what is the probability of producing two heads in three tosses? There are three possible sequences of three tosses in which heads is cast two times: heads-heads-tails, heads-tails-heads, and tails-heads-heads. The probability of each sequence is equal to the product of the probabilities of the single tosses, $\frac{1}{2} \cdot \frac{1}{2} \cdot \frac{1}{2} = \frac{1}{8}$. Therefore, the probability of casting two heads in 3 tosses is 3/8 (fig. 9.7).

Fig. 9.7. An example of Bernoulli trials: sequences obtainable tossing a coin three times or three coins at the same time. The probability that the relative frequency of heads is 2/3 is $3 \cdot (1/2)^3 = 3/8$.

The probability of 5 heads in 10 tosses, or of 100 heads in 500 tosses, or of any other frequency in any number of tosses can be likewise calculated. In this example, the sequences for a given number of trials are all equally likely; therefore, it is possible to calculate the probability of any frequency either as compound probabilities or as the ratios of favourable to total outcomes. A less elementary example is Bernoulli's urn containing tokens of two colours. Let us consider a simplified version, namely, an urn containing three white balls and two black balls. If the drawn ball is replaced in the urn every time, the draws constitute Bernoulli trials. In this case, the probabilities of drawing a white and a black ball are 3/5 and 2/5 respectively, and the probability of a sequence will depend on the numbers of white and black draws. The probability of a given frequency of successes (say, a certain number of black balls in a certain number of trials) can be calculated by multiplying the probabilities of the single draws and counting the number of sequences containing the desired number of successes (fig. 9.8).

As Bernoulli explained, the probability that the relative frequency is exactly equal to the probability of success decreases with the number of trials. It is less probable to obtain exactly 5 heads in 10 coin tosses than 1 heads in 2 tosses, and even less probable to

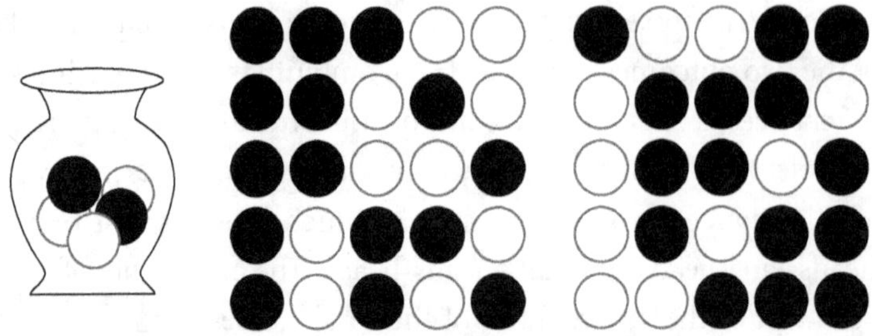

Fig. 9.8. Another example of Bernoulli trials: draws with replacement from an urn with 2 black and 3 white balls. In 5 trials ($n = 5$), the probability that the relative frequency of black balls is $k/n = 2/5$ is $10(2/5)^2(3/5)^3 \cong 0.35$. For $n = 10$ the probability of $k/n = 2/5$ drops to $210(2/5)^4(3/5)^6 \cong 0.25$. In this case, however, it is possible for the frequency to assume also other values close to 2/5, as for example 3/10 and 5/10, so that the probability of $3/10 \le k/n \le 5/10$ becomes 0.67.

obtain exactly 50 heads in 100 tosses, or 500 heads in 1000 tosses. As the number of trials increases, however, it becomes more probable that the relative frequency is *close to* the probability of success, for example that the relative frequency of heads is between 0.4 and 0.6, or 0.49 and 0.51, or 0.499 and 0.501. Bernoulli derived the general expression of the probability of any frequency of successes in any number of trials, given the probability of success in a single trial. He realized that for a given number of trials, n, and given the probabilities of success and failure, p and q, the probabilities of the frequencies were equal to the terms of the expansion of the nth power of a binomial, $(p+q)^n$, the coefficients of which are the binomial coefficients, that is, the terms of Pascal's triangle (see fig. 9.5 and footnote 5). Then he proved that the probability that the relative frequency was within a small distance from the probability of success became higher with the number of trials. Calling p the probability of a single success, ε any small number, k the number of successes and n the number of trials, Bernoulli proved that the probability that the frequency k/n would differ from p for less than ε increased with n and would become equal to 1 for infinite n. In symbols,

$$\Pr[p-\varepsilon<\frac{k}{n}<p+\varepsilon]\xrightarrow[n\to\infty]{}1$$

Had Bernoulli aimed for certainty, this result would have been discouraging, since it showed that certainty could never be reached: only after an *infinite* number of trials could you be certain that the observed relative frequency was close to p; however, he considered the theorem a means to calculate how many trials were required to reach moral certainty that the observed frequency approximated the unknown probability. Although conceptually clear, the problem was mathematically difficult to solve. Availing himself of simplifying approximations, he estimated that in the example of the urn, the number of draws required to reach a probability of 0.999 that the relative frequency was within 1/50 of the true ratio of white to black tokens was 25550. Modern commentators speculate that this number, an impossibly large number of repetitions of an experiment to carry out to determine anything, was the reason why Bernoulli was not happy with his result and abandoned the book. Later authors, as we shall see, developed more refined methods of calculation and showed that Bernoulli's number was about four times larger than necessary. Nevertheless, Bernoulli expressed no disappointment. On the contrary, his last words in the unfinished book were an upbeat reminder of his deterministic worldview (Bernoulli 2006, 339),

> Whence at last this remarkable result is seen to follow, that if the observations of all events were continued for the whole of eternity (with the probability finally transformed into perfect certainty) then everything in the world would be observed to happen in fixed ratios and with a constant law of alternation. Thus even in the most accidental and fortuitous we would be bound to acknowledge a certain quasi-necessity and, so to speak, fatality.

The Normal Curve

Jakob Bernoulli derived the general expression of the probability of any frequency in a series of n trials, given the probability of success in a single trial, p. In modern terms, he found the first *probability distribution* of a chance variable. Specifically, his chance variable was the frequency (or relative frequency), and its distribution is today called the binomial distribution. The difficulty that confounded him was that, in order to answer the question of how many trials were required to reach moral certainty that the relative frequency approximated p, it was necessary to sum up the probabilities of all the relative frequencies in an interval around p. Bernoulli resorted to crude approximations. A more refined method was devised by Abraham De Moivre, a French mathematician who emigrated to England in 1688 to escape the persecution of Protestants in his native country. De Moivre became interested in the mathematics of chances while in school, studying Huygens's treatise. Upon his arrival in England, he studied Newton's early version of calculus, the method of fluxions. He was then able to creatively combine the two, applying the techniques of infinite series and of integration to solve Bernoulli's problem. He published his results in *The Doctrine of Chances*, which went through three editions between 1718 and 1733. De Moivre's breakthrough consisted of finding that for large values of n the frequencies, which were discrete values, could be approximated by a continuous variable and the corresponding probabilities could be approximated by a bell-shaped curve that had its maximum point at p and was symmetrical around it (fig. 9.9). As he put it, writing at a time when graphing was still not a common way to visualize mathematical expressions,

> If all the terms of the binomial are thought of as set upright, equally spaced at right angles to and above the straight line, the extremities of the terms follow a curve. The curve so described has two inflection points, one on each side of the maximal term. (De Moivre quoted in Stigler 1986, 76–77.)

He was thus able to replace the impossibly difficult summation of the probabilities around the central point with the integral of the curve over an interval centered on p. The curve that he devised as an approximation of the binomial distribution became the

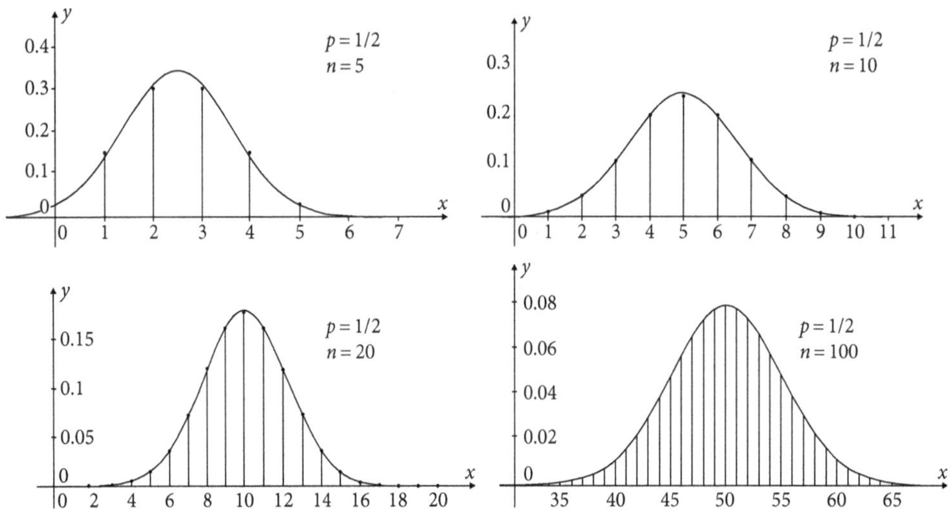

Fig. 9.9. Modern graphical representation of the probabilities of frequencies in Bernoulli trials (binomial distributions) with $p = 1/2$ for increasing numbers of trials, and their approximations by normal curves. With a different value of p the graphs for low values of n would not be symmetrical, but they would become increasingly symmetrical and closer to the corresponding normal curves as n increased.

central mathematical tool of probability theory. It is today known with various names, bell-shaped or bell curve, error curve, normal distribution, or Gaussian function.[9] The reason for the first name is self-evident; the reason for the second will become evident in the next section. The name "normal" became popular in the 1880s, after the use of the curve to represent the frequency of many quantities had become so widespread that it came to be thought of as a general norm, if not a law, of nature. And the name "Gaussian" is a classical example of Stigler's Law of Eponymy, as it derives from the German mathematician Carl Friedrich Gauss, who contributed decisively to demonstrating the potential of the curve but did not discover it (Stigler 1999).

De Moivre thus understood that for a sufficiently high number of trials, the normal curve could be used to determine the probability that the frequency would be within any given interval around the central value (in modern terms, the mean), and that the distance of the inflection points of the curve from the central value could be used as a measure of the dispersion of the frequencies (standard deviation). The probability that the relative frequency would be between the inflection points of the curve was approximately 0.683, and it became approximately 0.999 for an interval three times as large (fig. 9.10). As a specific example, De Moivre considered a number n of Bernoulli trials with $p=1/2$. In that case, the distance of the inflection points from the central value was equal to $\sqrt{n}/2$; therefore, a probability of 0.998 that the relative frequency differed from p by less than 0.025 was obtained for $n=3600$. Applied to Jakob Bernoulli's example, the normal curve method gives approximately 6500 trials.

The Theory of Errors and the Problem of Inverse Probabilities

Jakob Bernoulli and Abraham De Moivre opened the way to the use of mathematical probability to draw inferences on the basis of large numbers of observations. They hoped that the Law of Large Numbers would enable mathematicians to determine, probabilistically and a posteriori, the otherwise unknown probabilities of single events by observing the frequencies of the events in the long run. In De Moivre's words,

> From this it follows, that if after taking a great number of Experiments, it should be perceived that the happenings and failings have been nearly in a certain proportion, such as 2 to 1, it may safely be concluded that the

[9]The Russian Pafnuty Lvovich Chebichev clarified the concepts of random variable and probability distributions in the 1860s. Today, mathematicians talk of the normal, Gaussian, or error distribution rather than just "curve".

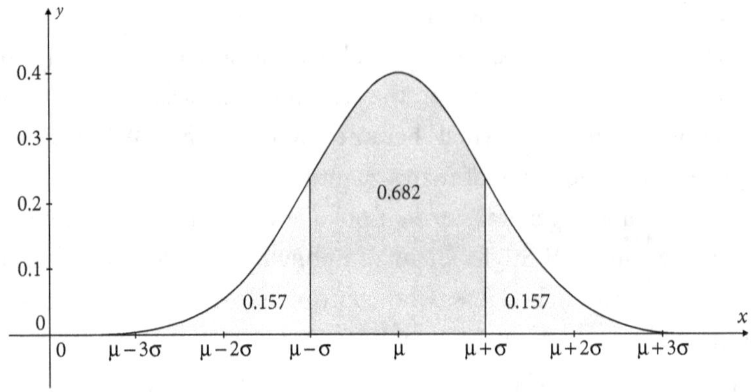

Fig. 9.10. The normal curve. The curve is symmetrical around the maximal point μ, the *mean*. The distance σ between μ and the inflection points (where the curve changes from convex to concave and vice versa) measures the dispersion of values around the mean, and it is called the *standard deviation*. The total area under the curve is 1; the area of the 1σ interval is 0.682, and the area of the 3σ interval is 0.997.

> Probabilities of happening or failing at any one time assigned will be very near in that proportion, and that the greater the number of Experiments has been, so much nearer the Truth will the conjectures be that are derived from them. (De Moivre quoted in Stigler 1986, 87.)

Laplace and a few others realized, however, that neither Bernoulli nor De Moivre had reached their stated goal of a safe method of conjecturing. Bernoulli's theorem was only a stepping stone toward a mathematical theory of inductive inference. Bernoulli's theorem allowed the calculation of the probability that, given p, the relative frequency k/n would be close to it. The goal, however, was to find the probability that, given an observed k/n, the unknown p would be close to it. The relation between p and k/n was a special case of cause-effect relation, where p was the cause and k/n the effect. More generally, considering any phenomenon that needed to be explained by an unknown cause, the goal of inductive inference was to determine the hidden cause from the observed effects. The Law of Large Numbers provided the probability of the effects given the cause, whereas what inductive inference required was the "inverse probability", the probability of the cause given the effect.

To see why these two probabilities are not the same let us consider the urn example in the case in which the ratio of white to black balls contained in the urn is unknown. Let us suppose that we perform 100 draws and observe approximately 40 black and 60 white balls. Can we conclude that the unknown ratio is 2/5 more probably than not? Clearly, not. The same empirical result could be produced also if the unknown ratio were different from that. If, for instance, the unknown ratio were 1/5, it would still be possible to draw 40 black balls in 100 trials. Can we conclude, perhaps, that the ratio of 1/5 (just as any other ratio) is indeed possible, yet less probable than 2/5? The answer is yes only if the two ratios, $p = 2/5$ and $p = 1/5$ are a priori equally likely, or if $p = 2/5$ is a priori more likely than $p = 1/5$. Imagine that the urn contained pearls instead of balls. In this case, a lower ratio of black to white draws would be a priori more likely than a higher ratio because black pearls are much rarer than white ones. Drawing 40 black pearls in 100 would then be more probably a lucky run than the result of an extraordinarily large (and expensive) number of black pearls in the urn. The problem of inverse probability cannot in general be reduced to simply equating the probability of the cause given the effect to the probability of the effect given the cause. Its solution has to combine the probabilities on the effects with some precondition of the probabilities of the causes, and it has to imply also a mechanism through which the probabilities of the causes are changed by the observations we make.

In the last part of the eighteenth century, first the Englishman Thomas Bayes and then, independently, Laplace developed a method to calculate inverse probabilities using a theorem that became known as Bayes's theorem. They defined the *conditional probability* an event, A, given another event, B (in symbols, Pr(A/B)), as the ratio between the probability of A and B occurring together ($A \cap B$) and the probability of B:

$$\Pr(A/B) = \frac{\Pr(A \cap B)}{\Pr(B)}$$

From this they derived the relation between the probability of two events that are identified as cause and effect, $C$and E, and the conditional probabilities of the cause given the effect and of the effect given the cause. In the simplest form, Bayes's theorem is:

$$\Pr(C/E) = \frac{\Pr(E/C) \cdot \Pr(C)}{\Pr(E)}$$

In principle, the inverse probability, $\Pr(C/E)$ could be calculated by means of this relation, provided that the direct probabilities, $\Pr(E/C)$ and $\Pr(E)$ were known. The major obstacle was that it required also $\Pr(C)$, the probability of the cause prior to the observation of the effect. This was and remains the main difficulty in the application of Bayes's theorem. Laplace simplified the problem by making the assumption that a complete set of possible causes was known, and that they were all equally likely. With this precondition, inverse probability becomes simply proportional to direct probability.

Pierre-Simon de Laplace was one of the most prominent among the eighteenth-century mathematicians who expanded and completed Newtonian mechanics using differential and integral calculus. His theoretical study of the motions of the planets in the solar system culminated in the publication of a monumental work in five volumes, *Traité de mécanique céléste* (*Treatise on Celestial Mechanics*, 1799–1825), for which he was hailed as "the French Newton". He served as a high-ranking civil servant and member of the Paris Academy of Sciences throughout the French Revolution, Napoleon's empire, and the restoration of the Bourbon monarchy. In his capacities as a member of the Academy of Sciences, the director of the Bureau of Longitudes, and the director of the Paris Astronomical Observatory, he took part in ambitious projects such as the sweeping reform of weights and measures that replaced the congeries of old units of measurement with an orderly system of standardized decimal units. Like all other professional mathematicians of his time, he practised mixed mathematics. Moreover, as a civil servant, he was especially involved with questions related to measurements. It was to deal with uncertainty in measurements that he became interested in inverse probability. He gave a comprehensive exposition of his results in probability theory in the treatise, *Théorie analytique des probabilities* (*Analytical Theory of Probabilities*, 1812), to which he added as a non-technical preface the *Essai philosophique sur les probabilities* (*Philosophical Essay on Probabilities*, 1814). In the *Essai*, he provided a famous statement of the deterministic view of the universe. He depicted a hypothetical superhuman "intellect", now famous as "Laplace's demon", capable of knowing the position and velocity of every particle in the universe and the forces acting on it, and capable of carrying about the necessary calculations to deduce every future position and velocity from the Newtonian laws of motion. To such an intellect, uncertainty would not exist, as past, present, and future would be completely known. Probability, Laplace maintained, was an intellectual tool made necessary by the limits of human knowledge. Nevertheless, he believed, as did his colleagues, that the uncertainty of a rational mind was quantifiable and amenable to rigorous computation. In his view, probability was "nothing more than common sense reduced to calculus." He envisioned the theory of probability as the mathematics of human life in the same way as calculus was the mathematics of the physical world.

Laplace arrived at his definition of inverse probability through the problem of dealing with errors of measurement in astronomical observations. The problem of measurement errors consisted of determining the true value of a physical quantity, as for example the position of planet at a given time, from the results of measurements. Astronomers, experimenters, and professionals who dealt with measurements knew that the results of repeated measurements of the same quantity did not coincide exactly but rather tended to cluster, presumably, around the true value. Once every conceivable precaution had been taken to eliminate systematic biases in the measuring instruments and techniques, and every detectable mistake had been discarded, the results were spread over a range of values, leaving open the question of where exactly within that range the true value lay. It had become common practice among astronomers to choose the average or mean of a set of good measurements as the best estimate of the true value. But was there a way to justify this choice mathematically? Laplace applied probabilistic reasoning to the problem. He treated the error of a measurement, that is, the difference between the actual measurement and the true value, as a chance variable with a predetermined distribution of probabilities. The probability distribution of errors had to satisfy three conditions. First, it had to be symmetrical around zero, since in absence of systematic biases there had to be equal probability that the observed value was lower or higher than the true value. Second, the probability of any finite error had to be larger than zero and to become indefinitely smaller as the error increased; in other terms, the probability curve had to be asymptotic to the axis representing the errors. Finally, the total area under the curve had to be equal to 1. But these conditions were insufficient to identify exactly what curve represented the probability of errors, and Laplace tried several possibilities before finding the right answer. Gauss, who was also deeply involved in celestial mechanics, as well as astronomical and geodesic measurements, was the first to argue that the correct curve to represent the error distribution was the normal curve. He showed that if the errors were distributed according to De Moivre's curve, then taking the mean of a set of measurements as the best estimate of the true value was mathematically justifiable, as it was the value that minimized the expected errors. Laplace endorsed Gauss's proposal and supplied a stronger justification for it. Generalizing De Moivre's demonstration, he showed that not only the frequency of successes in n trials, but also any sum of n terms was distributed approximately according to De Moivre's curve around the mean value. This theorem, known as the "Central Limit Theorem", allowed Laplace to show that if the mean was chosen as the result of a series of measurements, the error of the mean followed, at least approximately, De Moivre's curve. Furthermore, since every error could be conceived as the sum of a number of tiny deviations from the true value, it also followed approximately De Moivre's curve (hypothesis of elementary errors). Thanks to the work of Laplace and Gauss, the curve now famous as the Gaussian or normal curve became the centerpiece of the probabilistic theory of errors, which became the most successful of the early applications of probability theory.

Nevertheless, many of Laplace's successors criticized and ultimately rejected Laplace's inverse probability because they saw his precondition of equally likely causes as a blind presumption based only on ignorance. The majority of scientists in the nineteenth century subscribed to a philosophy of science according to which scientific knowledge must rest exclusively on empirical bases. Consequently, they discarded Bernoulli and Laplace's epistemic definition of probability as constituting a "subjective" interpretation. In their view, the only sufficiently empirical definition of probability was that of relative frequency in the long run—the "frequentist" definition. Frequency probabilities could only be assigned to the outcomes of repeatable experiments, or to events that could be regarded as replications of the same experiment many times over. The frequentists denied that probabilities could be assigned to causes or hypotheses, for causes and hypotheses did not in general possess the attribute of repeatability. For this reason, they sought a theory of probabilistic inference that did not involve inverse probabilities. In the early part of the twentieth century, mathematical statisticians devised methods to test precisely formulated hypotheses by means of carefully conducted experiments using solely the direct probability of experimental results given the hypotheses. These methods do not permit the calculation of the probability of the hypothesis—although they are often misunderstood in this way—but they provide the means to reject or accept the hypothesis on the basis of the experimental results. Epistemic probabilities were revitalized in the second part of the twentieth century, and Bayes's theorem was then re-interpreted as a technique to update the probabilities of causes on the basis of observations. It remains to this day a matter of contention between different schools of thought on probability.

9.4 From Classical to Statistical Probability

For the classical probabilists from Jakob Bernoulli to Laplace, probability was a composite concept. Part of it coincided with chance in any process that, thanks to some physical symmetry, presented a finite set of possible outcomes which were all equally favoured by the laws of nature and hence equally likely to occur. Let us call this component *aleatory* probability, from the Latin word *alea*, which means die and was often used by the early calculators of chances as a synonym for chance. Closely associated with aleatory chances was the component of probability that referred to the frequency of occurrence of a given event in a large number of repeated experiments, as in a long series of tosses of a coin, or in a large number of occurrences that could be regarded as repetitions of the same experiment, as in the toss of a sack-full of coins. This is the *frequentist* component of probability. These two components are sometimes referred to together as *objective* probability, because they refer to properties of the objects of knowledge, or to the state of the world independently of our knowledge. The third component referred to the weight of evidence and testimony for or against an uncertain hypothesis. We could perhaps call this "forensic" or "argumentative" probability. The fourth component indicated the degree of rational belief, or the numerical measure of the partial belief that a rational "man" (today, with gender-neutral language, we would say a rational person) would place on an uncertain hypothesis. Let us call this *epistemic* probability, as it concerns an aspect of our state of knowledge.

During the first period of the history of the theory of probability, the classical period, the Newtonian view of the world, the belief in universal reason, and specific theories of knowledge that were popular during the Enlightenment supported the fusion of the four components of probability into a single, if composite, concept. By the mid-nineteenth century, however, mathematicians saw a conflict between the epistemic and non-epistemic aspects, and drew a distinction between subjective and objective interpretations of the quantity called probability. While the deterministic view of the physical world continued to dominate, the French Revolution, the Napoleonic Wars, and the social and political turmoil of the Industrial Age drastically weakened the faith in one universal "rational man".[10] Uncertainty and belief began to be seen as matters of personal perception and intuition,

[10] Physical determinism was slowly undermined in the late nineteenth century as well, and it was overthrown by Quantum Mechanics in the early decades of the twentieth century.

too erratic and obscure to be quantified, and therefore unsuitable to support a mathematical theory. The view of science that prevailed was the positivist view, according to which the rigorous exercise of the correct scientific method would enable well-trained and disciplined researchers to ascertain hard, neutral facts, and thereby to produce knowledge that was independent of any state of mind of the knower. The majority of mathematicians after Laplace adopted the positivist standpoint and distanced themselves from the classical views of probability. Aiming to cleanse the theory of any suspicion of subjectivity, they re-defined probability exclusively as relative frequency in the long run. The nineteenth century saw the rise of statistics, the study of populations by means of the collection of numerical data. Statistical data were seen as empirical facts as hard and neutral as the data of physics experiments, and as practically perfect instantiations of the Law of Large Numbers. Statistics offered the best field of application for the calculus of probabilities. For this reason, the second period of the history of probability theory can be characterized as the phase of "statistical probability".

Sources

Bernoulli, Jakob. *The Art of Conjecturing, together with Letter to a Friend on Sets in Court Tennis*, edited by Edith Dudley Sylla. Baltimore: Johns Hopkins University Press, 2006.

Daston, Lorraine. *Classical Probability in the Enlightenment*. Princeton: Princeton University Press, 1989.

Gigerenzer, Gerd et al. *The Empire of Chance. How Probability Changed Science and Everyday Life*. Cambridge: Cambridge University Press, 1989.

Hacking, Ian. *The Emergence of Probability*. Cambridge: Cambridge University Press, 1975.

Hacking, Ian. *Probability and Inductive Logic*. Cambridge: Cambridge University Press, 2001.

Ore, Øystein. "Pascal and the Invention of Probability Theory", *American Mathematical Monthly* 67 (1960), 409–419.

Ore, Øystein. *Cardano, The Gambling Scholar* (New York: Dover Publications, 1965).

Stigler, Stephen M. *The History of Statistics. The Measurement of Uncertainty before 1900*. Cambridge, MA and London, England: The Harvard University Press, 1986.

Stigler, Stephen M. "Stigler's Law of Eponymy". Reprinted in *Statistics on the Table. The History of Statistical Concepts and Methods*, 277–290. Cambridge, MA and London, England: Harvard University Press, 1999.

Appendix

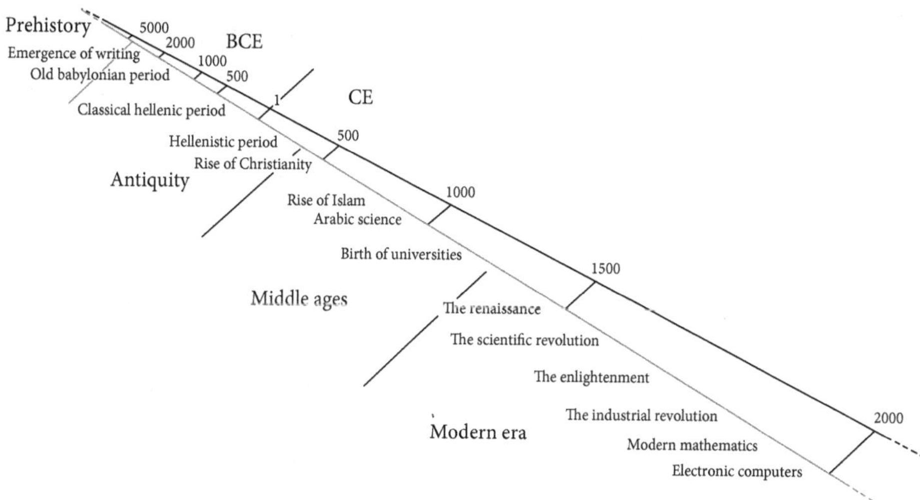
Prehistory
5000
2000
BCE
Emergence of writing
1000
Old babylonian period
500
1
CE
Classical hellenic period
500
Hellenistic period
Rise of Christianity
Antiquity
Rise of Islam
1000
Arabic science
Birth of universities
1500
Middle ages
The renaissance
The scientific revolution
The enlightenment
The industrial revolution
2000
Modern era
Modern mathematics
Electronic computers

www.ingramcontent.com/pod-product-compliance
Lightning Source LLC
LaVergne TN
LVHW080049210926
848481LV00011B/3311